WILL POPULATION +
TECHNOLOGY =
ARMAGEDDON?

HENRY T. MULLINS, PhD

Cover: NASA image of nuclear explosion.

Printed by Create Space (www.createspace.com)
(#3376991) – September, 2009

Revised January, 2010

First of a series of educational/environmental books
offered by *Environmental Earth Systems, Inc.*
(www.environmentalearthsystems.com)

PREFACE

Why did I write this book? I believe this topic to be the most significant scientific issue of our time for the long-term survival of our species! Planet Earth's global population is on a very rapid rise and we have recently made great advances in technology. Some of this technology is "good", some is "bad", and some is "useless". The combination of excessive human population and technology gone awry is a potentially highly combustible combination!

I do not know the precise answer to the question that is posed as the title of this book. Thus, one may legitimately ask – "well then, why write the book?" The purpose for writing this book is to attempt to get our global society to seriously contemplate this ultimate question, stimulate scientifically based intellectual discussion, disseminate scientific literacy, and provide a glimmer of hope.

This book is based on science but it is written for the educated general public, policy makers, high school students, and introductory level college students. Thus, you do not need to be a scientist to read this book. This book also is not a "doomsday book" like Stanford University biology professor Dr. Paul Ehrlich's 1968 book *The Population Bomb* in which he predicted that hundreds of millions of people would die by the end of the 1980s due to overpopulation and limited resources - similar to Thomas Malthus' (1766-1834) "catastrophe theory". This book is also different from The Club of Rome think tank's 1970s reports – *Limits to Growth* and *Mankind at the Turning Point* – that argued that economic growth can not be indefinite because of limited natural resources such as oil. *Limits to Growth* is the best selling environmental book in history with over 30 million copies sold.

I also want to assure you that I do not walk the streets of my hometown with a sandwich board proclaiming that "the end is near". This book is a well thought out, scientific presentation on the long-term survival of our species based on factual data on human population and technology at the present time. These data are then integrated with, and written from, a relatively new view of Planet Earth referred to as *Earth System Science* that will be defined in the body of the book. And I do offer a possible alternative to "business as usual". My ultimate goal is for you to find this book interesting and intellectually stimulating – it is that simple. I have no guaranteed answers, I do not profess to have any guaranteed solutions, and I make no predictions.

Someone asked me the other day, how long has it taken me to prepare to write this book? After pondering the question for a few seconds, I responded "unknowingly for forty years - knowingly for about a year and a half". In other words, the preparation for this book is a culmination of my formal scientific education and being a university professor for 30+ years.

As such, I would like to thank all my former professors, professional colleagues, peers, and graduate students that have helped to shape my way of thinking about our planet and our species over the past four decades. I also want to thank Syracuse University for allowing me the dedicated time to write this book while on sabbatical leave at Colgate University in the spring of 2009. And finally, I wish to acknowledge the critical reviews of this book by Dr. Adam Burnett in the Geography Department at Colgate University and Dr. Stephen W. Link, Professor of Psychology at Brookhaven College in Dallas, Texas. I would also like to thank Dr. Burnett and Mr. Joe Eakin at Colgate University, Mr. Peter Cattaneo at Syracuse University, and my son Patrick Mullins for computer assistance, as well as Mr. Aaron Diefendorf at Penn State University for computer graphics. I dedicate this book to all these wonderful people and institutions, as well as my three children – Patrick, Katie and Caroline.

ABOUT THE AUTHOR

My paternal roots are in New York City (lower Manhattan) but I grew up in the small town of Ghent, New York and graduated from Chatham High School in 1969 – my mother's hometown. I received my Bachelor of Science Degree with honors in geology from The State University of New York at Oneonta in 1973. I then earned my Masters of Science degree in marine geology at Duke University in 1975 before moving down tobacco road to the University of North Carolina at Chapel Hill for my PhD in oceanography that was completed in May 1978.

In September of 1978 I began my first academic job as an assistant professor of oceanography at Moss Landing Marine Laboratories of the California State University System located on Monterey Bay. It was a great five years of my life conducting marine geological and geophysical research offshore central California and teaching a variety of courses in geological oceanography. I earned promotion to associate professor and was granted tenure in 1982.

On the first of January, 1983 I began a 25+ year career as a professor of geology and earth sciences at Syracuse University, from where I will retire in June 2011. I have been a full professor at Syracuse University since 1990. While at Syracuse I conducted research on a variety of topics in a myriad of places around the world. I spent more than three decades studying the development of carbonate platforms in the Florida-Bahamas-Caribbean area and was fortunate to visit the bottom of the ocean at a depth of ~12,000 feet in the Deep Submergence Research Vessel (DSRV) ALVIN that is probably best known for its role in the discovery of *The Titanic*. I have spent many 100s of days out at sea as chief scientist on various research vessels, including the maiden voyage of the Deep Sea Drilling vessel JOIDES RESOLUTION, and oversaw research funding in excess of one million dollars.

When I became the father of three wonderful children (Patrick, Katie, and Caroline) I intentionally changed my research area to the nearby Finger Lakes of central New York State so that I could be closer to home while maintaining my expertise in conducting research in water covered areas. Over the past two decades I have been involved in the origin of large lakes such as the Finger Lakes as well as those in southern British Columbia, Canada. My research interests have since evolved into the lake-sediment record of climate and environmental change over geologic time (particularly the last 11,000 years) that ultimately brought me to lakes in the land of my heritage – Ireland. I have also become interested in the impact that humans have on the Earth System, and what impacts the Earth System may have on humans, which has stimulated me to write this book.

I had the pleasure and honor of serving as editor of the scientific journal GEOLOGY (published by the *Geological Society of America* in Boulder, Colorado) from 1990-1995, which is arguably the most prestigious geo-science journal in the world. I and my co-authors have also published over 100 scientific articles in peer reviewed international journals since 1977. I am a Fellow of the Geological Society of America, and a member of the American Geophysical Union. I was proud to receive the *Geological Society of America's* Distinguished Service Medal in 1995 as well as Syracuse University's William Wasserstrom Award for Excellence In Graduate Teaching in 1997.

While at Syracuse University I taught a variety of courses from large introductory level classes to supervising numerous Masters and PhD students, some of whom are now tenured professors at other academic institutions. At the present time my teaching interests include introductory courses in Earth System Science, Oceanography, and a graduate level course in the Geologic Record of Global Change. In all my courses, societal relevance is always strongly integrated into my teaching.

TABLE OF CONTENTS

III) POPULATION

Who were the first two people on Earth? Some deeply religious people may quickly answer "Adam & Eve". Others with a more scientific perspective might say they do not know, but that humans evolved from more primitive beings as proposed by Darwin. And those that do not think deeply about such questions might simply say "I don't care". The bottom line, however, is that it does not matter who the two first two humans were, because the fact of the matter is that we are here, and we have very rapidly become the most dominant species on Planet Earth!

It is unlikely that we will ever know the answer to the question posed above, but what is certain is that the Earth System provided early humans with an enormous bounty of food, water, habitat and an appropriate climate. All that any species has to do to survive is to find food, successfully reproduce, and avoid predators. We also know that the Earth is about 4.6 billion years old (based on radiometric dating of Earth rocks, Moon rocks, and meteorites) and that life of one form or another has existed on our planet for at least the past 3.8 billion years (based on the fossil record).

The Earth System: The concept of Planet Earth as an integrated, holistic system that has the ability to "self-regulate" its environment has its roots in the Gaia Theory of Dr. James Lovelock that emerged during the 1960s and 1970s when Lovelock was a NASA scientist. From its inception, Gaia (Greek Goddess of Mother Earth) was challenged by the established scientific community. It has since evolved into theory that many of us believe is the wave of the future toward a full knowledge and understanding of our planet. The textbook - "The Earth System" - by Dr. Lee Kump and his colleagues at the Pennsylvanian State University, is now widely used to teach introductory Earth Science at universities and colleges around the world.

The Earth System consists of four fundamental components – the solid Earth (lithosphere), the water Earth (hydrosphere), the gaseous Earth (atmosphere), and the living Earth (biosphere) that freely interact, in a coupled fashion, such that it is capable of "self-regulation" – the ability of our global environment to remain essentially the same over time. A key ingredient to Earth System Science is the realization that "the whole is greater than the simple sum of its parts". There is roundtrip flow of energy, matter and information between the major components via "couplings" and "feedbacks" that do not require intelligence and are very much Darwinian in concept. Positive feedbacks tend to amplify a change in the

Earth System, whereas negative feedbacks tend to dampen, or limit, disturbances.

When we stand outside our feet are on the lithosphere and our heads are in the atmosphere. The lithosphere is very dynamic with its myriad of geological processes controlled by plate tectonics over long periods of time (millions of years). So to is the atmosphere which consists of 78% nitrogen, 21% oxygen, and less than 1% greenhouse gases, but operates on a much shorter time scale (daily to thousands of years) than the lithosphere. When we swim in a lake or ocean, we enter the hydrosphere where 97% of Earth's water is retained in the world ocean, 2% in ice sheets (largely Greenland and Antarctica), and only 1% as freshwater on or in land that operates at intermediate time scales of decades, to centuries to millennia. We, as living organisms are part of the biosphere, along with 10s of millions of other species ranging from the smallest bacteria to the largest creature to ever inhabit Earth – the blue whale. Some scientists, such as Dr. William Ruddiman at the University of Virginia argue that humans have been perturbing the Earth System for at least the past 5,000 to 8,000 years when we evolved from being hunter-gathers, to an agrarian (agricultural) global society, and in so doing started to "manage" and exploit the Earth. Such management exploded around

1850 with the Industrial Revolution, and again in the 1940s following World War II - both due to large-scale increases and advancements in technology.

Mass of Humanity: Human evolution has occurred in distinct stages or steps over the past ~ 5 million years with each step resulting in increased mobility, ability to use tools and an increase in brain size. About 2.5 million years ago the genus *Homo* evolved as a lean, graceful creature with a larger head and braincase than its predecessors. Increasing brain size is a critical component of our intellectual capacity, as is "folding in". In fact, over the past 3-4 million years the volume of our ancestors' braincase has tripled. Modern humans, *Homo sapiens* – or "intelligent man" - first appeared on Earth sometime between 200,000 and 100,000 years ago. Overall, human evolution has been rapid, and it is very important to realize that if *Homo sapiens* have been on Earth for a maximum of 200,000 years that "intelligent man" has existed for less than 0.0005 % of all of geologic time – an amazingly short interval considering our intellectual capabilities and dominance as a species on Planet Earth.

Because of our intelligence, we are very well adapted to the basic requirements of population growth – and populate we have! About 11,000 years ago, when Earth entered our present relatively warm interglacial period, it is estimated that early humans had multiplied into somewhere between one and ten million people. By the first year AD this number increased to approximately 300 million people. We topped the one billion mark by 1800, and more than doubled that to about 2.5 billion people in the year that I was born (1951). At present (2009), Earth is home to some 6.5 billion people, and projections for the year 2050 range from a staggering 8 to 12 billion!

The global human population growth curve over time is "J-shaped" and became exponential (increases by powers greater than one) with the onset of the scientific, medical and industrial revolutions of the late 1800s and 1900s. In order to understand, in simple terms, how global human population can grow so rapidly, let's start out with a real life example and then make some simplified assumptions.

My paternal grandparents immigrated to the United States from County Cork, Ireland in the late 1800s. They met, fell in love and proceeded to have 14 children in lower Manhattan. Now, back in Ireland in those days a large family was a priority so as to ensure enough workers on the farm – but in New York City? So, within a single generation two people multiplied into 16 humans. Now, let's assume that all 14 of my aunts and uncles survived, married, and produced 10

Human Population (billions)

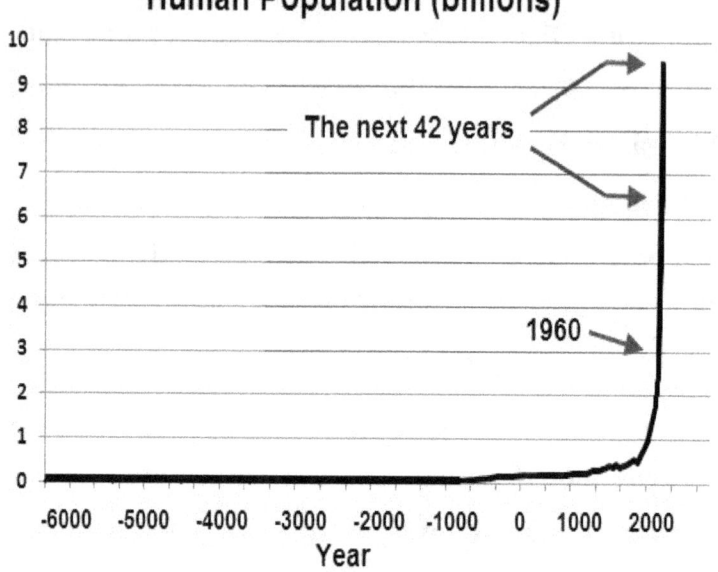

children each. At this point in time, what started out with two people (my grandparents) are now 140 people. Again, let's assume that all my relatives survive and each has yet an additional 10 children. Now we have 1400 people that began as two. And on and on it goes over time. As long as the global birth rate exceeds the global death rate, there will be net population growth at a potentially exponential pace.

Carrying Capacity: There are, however, limits to population growth of any organism, in any ecosystem. Scientists refer to this sustainable limit as the "carrying capacity" of an ecosystem. And, the Earth System is no different when it comes to humans –

there *is* a limited carrying capacity. What that magic number of people might be is not precisely known, nor when a threshold or "tipping point" within the Earth System might be crossed in time – but it will occur unless global population growth not only slows, but begins to decline. Some population scientists, such as the late Dr. Julian Simon (University of Maryland) dismiss the concept of a human carrying capacity completely on the grounds that technology will be able to solve the issue via large scale geo-engineering and management of Earth. And, although there is considerable uncertainty and lack of consensus, most population biologists estimate that the human carrying capacity of Planet Earth is somewhere between 10 and 20 billion people. We are presently at ~ 6.5 billion, and 10 billion is in the middle of the projected 8 to 12 billion people in a mere 41 years (2050)!

To further understand the concept of carrying capacity, I will use a simple analogy. I am a fisherman, so let's envision a small pond on a farmer's lot of land. The farmer wants to stock the pond with largemouth bass so as to be able to fish for recreational purposes. The farmer is not familiar with the concept of carrying capacity and he arbitrarily stocks the pond with 250 largemouth bass. Unknown to the farmer is that the natural carrying capacity for largemouth bass in this small pond is 500. A year latter the farmer goes to the pond and catches a number of nice sized bass that are released back into the pond. The following summer the farmer fishes again, and much to his pleasant surprise finds that the fishing has improved. He is able to catch more, larger bass in the same amount of time with the same amount of effort. The reason for this is that the farmer stocked 250 bass in a pond with a carrying capacity of 500. The bass had flourished, grown in size and multiplied in numbers.

The following summer the farmer returns to the pond and much to his amazement he finds that the fishing is even better – more, larger fish. During the following few summers, however, the farmer begins to notice that the fishing has become rather constant (stasis). The reason for this is that the pond has reached its natural carrying capacity of 500. So, the farmer decides to stock an additional 250 bass in the pond. The farmer is disappointed because the fishing does not improve - the pond "system" has self-regulated the bass population via negative feedbacks. Out of frustration, the farmer then decides to stock an additional 1000 bass in the pond. When he returns the following summer he finds that all the bass have died. Why? Because the farmer unknowingly overstocked the pond and greatly exceeded the pond's natural carrying capacity. There were so many bass in the pond that they ran out of food, started to die, and their decaying bodies used up all the oxygen in the pond and the remaining bass died of asphyxiation. In other words, despite the ability of the farm pond to self regulate a small excess in its carrying capacity, the large stocking event resulted in total collapse of the bass population.

Human Statistics: Now let's look at some statistical data about human population on Planet Earth – a much larger system than the farm pond discussed above. The United Nations estimated that global human population reached 6 billion people in October of 1999. According to the United States Census Bureau, U.S. population in April 2000 was ~ 281.4 million or ~ 5% of the global population. China had the largest percentage of the world's population at ~ 21%, followed by India with ~ 17 %.

To control its population China has taken a draconian approach referred to as the "one child policy" as part of its family planning policy. This policy officially restricts the number of children urban couples may have to one, although there are exemptions in rural areas and for ethnic minorities. Approximately 39.5 % of China' population is affected by this policy. The policy was designed to lessen social, economic and environmental problems in China, and since its inception in 1979 it is estimated that China has limited its population by about 300 million people. Over 75 % of the Chinese people support this policy. The "side effects" of this one-child policy are a disdain for female infants, abortion, neglect, abandonment and even infanticide. It also has changed China's gender demographics from the mean global value of 105 male births for every 100 females, to 114 males for every 100 females. China's fertility rate is now 1.7 births per woman, versus 2.1 in the United States. China's National Population and Family Planning Commission has stated that the "one child" policy will remain in effect for at least another decade.

Dr. C.R. Pope at Brigham Young University recently reported (*New England Journal of Medicine*) that people living in the United States are now living ~ 3 years longer than they did only two decades ago; ~ 2.5 years due to medical technology and ~ 0.5 years due to technological advancements in cleaner air. The United States began the year 2009 with a population of ~ 305.5 million people which is an increase of ~ 2.7 million people since the start of 2008 and an ~ 8 % increase since 2000. Some unknown, but significant portion of this increase is surely due to immigration, both legal and illegal.

The U.S. Central Intelligence Agency estimates that the 2008 fertility rate in the United States is equal to 2.1 children born per woman. However, Dr. Joel Cohen at Rockefeller and Columbia Universities and a member of the National Academy of Science, calculated that the *global* fertility rate is ~ 2.65 children per woman that will result in a growth rate of ~ 1.2 % per year, or about an additional 75 million people on Planet Earth each year.

The U.S. Census Bureau also estimates that in 2009 one birth is expected to occur every 8 seconds versus one death every 12 seconds. If correct, simple mathematics tells us that in 2009 the population of the United States will increase by ~ 1.3 million people independent of any immigration. If this rate remains constant, U.S. population will increase by 13 million people in 10 years and 130 million people in 100 years.

We must realize, however, that births per 1000 population vary considerably from country to country for socioeconomic reasons. According to the CIA, Niger has the highest number of births (49.62) per 1000 people and Japan the lowest at 7.87. The United States has a rate of 14.18 births per 1000 people. Thus, the population growth in Japan will be about half that of the United States while Niger's population will increase at a rate of ~ 3.5 times that of the U.S. A key point to bear in mind is that the population of all

21

three countries (Niger, Japan, and the United States) will *increase* but at variable rates, provided that death rates remain constant. What is important for the Earth System is what will be the *global* rate of population growth? As mentioned earlier, we presently have ~ 6.5 billion people on Earth and that is estimated to rise to 8 to 12 billion people in less than 50 years. That is equal to a global population increase of between 19 % and 46 % by the year 2050.

There are multiple reasons as to why the human population curve for Earth over time is "J-shaped". Lower rates of infant mortality, a decreased death rate, a more efficient global agricultural system, improved sanitary and dietary conditions, improved medical care and fertility clinics (technology) have all helped to expand our population. The average human lifespan prior to the modern era is estimated to have been on the order of only 25-30 years. In 1900, 50% of the people living in the United States could expect to live 50 years. In contrast, by 2000 the human lifespan in the United States increased more than 50% to 77 years. But bear in mind that the United States is a very affluent country - this increase does not apply equally to all countries of our global society for socioeconomic reasons.

Although there are little long-term data on infant mortality rates, and rates vary greatly around the world, consider that in Sweden the infant mortality rate decreased from about 30 deaths per 1000 live births in 1750 to only 11 in 2000. Also, in the United States infant mortality decreased 45% between 1980 and 2000, from 12.6 to 6.9 deaths per 1000 live births – both due primarily to advancements in medical technology.

Systematic agricultural practices can be traced back to about 9500 BC in the Middle East region (Iraq, Syria, Israel) when work was done by hand with very primitive tools. During the Middle Ages technology advanced to the point that irrigation systems, crop rotation and the plough were in use in North Africa and the Near East. Global crop exchange began after 1492 once Columbus "discovered" America. By the early 1800s crop yields per unit area of land were many times greater than that of the Middle Ages. And, with the technological advancements of tractors, fertilizers, and pesticides in the late 19th and 20th centuries, agricultural productivity blossomed, and as a consequence so did global food supplies. In 2005, China led the world in agricultural output followed by the European Union, India and the United States.

Models: Mathematicians have considered human population growth in quantitative terms that range from very simple to very complex. In simple form: birth rate – death rate = population growth, which is the simple model we will use for this book. In more complex form, two models have emerged – the "natural growth model"and the "coalition growth model"

Both of the complex mathematical models are consistent with Earth System Science concepts in that they implicitly include negative or positive feedbacks. In the natural growth model, natural processes are assumed to ultimately maintain a stable human population at Earth's carrying capacity via negative feedbacks – for example disease or war. The coalition growth model, however, allows for human population to exceed the carrying capacity of the Earth via positive feedbacks – for example increased technology. The natural growth model implies that human population will, at some time, reach an equilibrium or "steady state" (stasis – no change), whereas the coalition model implies unlimited human population growth with the potential for total collapse of human population. The key question is – what model (if either) will prevail in the potentially not too distant future? Only time will tell because we cannot predict the behavior of global society, and we do not yet completely understand all the possible feedbacks (positive or negative) within the Earth System.

IV) TECHNOLOGY

What is technology? Is it a computer, the space shuttle, an iPod, a flat screen HD television, the Internet? Or could it be more mundane things like cars, chain saws, and light bulbs? Or could it be very simple items such as the wheel, an axe, or a shovel? Or, is it all of these things and more?

Webster's dictionary provides us with three definitions of technology: 1) "the practical application of knowledge in a particular area" such as medical technology; 2) "a manner of accomplishing a task using technical processes" such as information technology; and, 3) "the specialized aspect of a particular field of endeavor" such as educational technology. Rather than memorizing all these definitions, let's simplify, for the purpose of this book, "technology" to mean some sort of "tool". The Internet is a tool to communicate or access information; a car is a tool to transport us from point A to point B; and, an axe is a tool to cut down trees.

Jane Goodall: This simple concept of technology as a "tool" is consistent with the groundbreaking work of the famous primatologist and anthropologist Jane Goodall who spent 45 years living with our primitive ancestors (chimpanzees) in the wild of Africa. Born in London, England on April 3, 1934 Dr. Goodall was given a lifelike chimpanzee toy by her father, which she named Jubilee that began her life long "love affair" of chimpanzees.

Goodall was invited to Africa by the famous anthropologist Dr. Louis Leaky to help his "dig" in Olduvai Gorge in eastern Africa. He encouraged Goodall to study the chimpanzees of Gombe Stream National Reserve, and she became known as one of "Leaky's Angels". Her research focus was to evaluate the differentiation between chimpanzees and Hominids (humans). Goodall's most significant discovery, and the importance to this book, was the observation that chimpanzees not only used tools ("technology") but also *made* tools that forced the scientific community of the time to reconsider the definition of what it means to be human.

Based on her studies Goodall stated that "I didn't see aggression to start with. There's no question (though) that chimpanzees become more aggressive as a result of crowding, as a result of competition for food". Her conclusions about crowding (overpopulation), making of tools (technology) and the aggressive behavior of our evolutionary ancestors is very relevant to this book. It shows that animals (be they chimpanzees or humans) become more aggressive with overpopulation, searching for food and the drive to survive. So, if we use Goodall's conclusions about chimpanzees as a simple analog, we can expect that in the future humans will become more aggressive as we become crowded together and compete for finite resources. I had the pleasure of listening to Dr. Goodall speak at Syracuse University a few years ago, and I can assure you that she is a very intelligent and inspiring person.

History: So, if we accept the simple definition of technology as some type of tool, we need to evaluate the role that technology has had on human population, as well as the role that human population has had on technology. There is an enormous, if not overwhelming, amount of information on the history of technology to be found in the scientific literature and/or on the Internet. I will attempt to focus on those parts of the history of technology that are most significant to this book – human population growth and the potential for Armageddon.

Technology is a method of creating new tools and products of those tools, which is a fundamental characteristic of our species. Although we do not possess highly instinctive reactions as many other species do, we have the unique ability to think systematically and creatively. "Man" is a toolmaker, and thus the development of technology is dependent on our evolution. Bear in mind, though, that there are also inherited social and cultural consequences of "too much" technology, such as a means to develop and use military power.

Historically, American sociologists and anthropologists such as Lewis H. Morgan (1818-1881), Leslie White (1900-1975), and Gerhard Lenski born in 1924 and a Professor Emeritus at the University of North Carolina, Chapel Hill theorized that technological progress has been the primary driving force of human civilization. Social evolution from "savagery", to "barbarism" to "civilization" can all be linked to technological advancements such as fire, domestication of animals, agriculture, metalworking, the alphabet and writing. Leslie White argued that "the primary function of culture was to harness and control energy". This energy emerged in five stages: 1) humans' own personal energy; 2) the energy of domesticated animals; 3) the energy of plants during the agricultural revolution; 4) the energy of natural resources such as coal, oil and natural gas; and, 5) nuclear energy.

Lenski, however, focused his attention on information. The more information a society had, especially that which allowed for managing its natural environment, the more advanced it was. He suggested four stages of development: 1) information passed on genetically; 2) information transferred by experience;

3) information translated via signs and logic; and, 4) information passed on by language and writing. Lenski argued that technological advancement of communication resulted in the development of economic and political systems that allowed human societies to evolve from hunter-gathers, to agrarians, and finally to our industrial global society.

Rather than chronologically list all of our technological advancements I will try to focus on some of the major "keystone" technologies that allowed our global population to grow, develop and prosper. Technology emerged during the Stone Age with the development of stone tools, fire, weapons and clothing that allowed for the transition of humans as nomads to settlers with rudimentary agriculture. During the "Metal Ages" technology advanced, and once humans invented axes, and were able to control fire, deforestation began on a limited basis. It also allowed for the development of stronger and more sophisticated weapons.

Widespread, organized agriculture began with the invention of the wooden plough about 8,000 BC in China that allowed for increased human survival. The wheel, often consider the greatest invention of all time, was developed about 4000 BC in Iran and greatly aided both agriculture and military operations. Ancient civilizations such as the Egyptians, Chinese,

Greeks, Romans, and Mayans pursued technological advances including machines, irrigation systems, seafaring equipment, gunpowder, natural gas, and concrete. The Muslim and South Asian Agricultural Revolutions resulted in the globalization of food crops and the industrial use of hydropower, tidal power, wind power and petroleum as well as explosive rockets and incendiary devices. Military technology advanced with the development of plate armor, steel crossbows, and the cannon/artillery.

The Industrial Revolution of the 1800s was fueled by "cheap energy" – coal. The steam engine was invented and manufacturing, mining, and metallurgy flourished. There were "amazing" developments in transportation, construction and communication, as well as advances in chemical, electrical, petroleum and steel technologies. This era also witnessed the development of the modern scientific method and approach that opened a new arena for technological advancements.

Many of these advancements, though, were tied to military research and development. The radio, radar, electronic computing, the telephone and nuclear power became available for military use. The National Academy of Engineers has ranked the most important technological advancements of the 20th century, some of which are electrification, the automobile, airplanes, mechanized agriculture, refrigeration, spacecraft, the Internet, household appliances, petroleum technology, lasers/fiber optics, nuclear technology, and material science. During the 21st century we can expect technology to develop even more rapidly in areas such as medical technology, nanotechnology, bioengineering, nuclear fusion, alternative energy, and superconductivity. We have come a long way from our initial use of stones as tools, but has all this technology been good for the long-term survival of our species?

Anthropocene: Geologically we live in the relatively warm, interglacial time interval of the last ~ 11,000 years known as the Holocene. Because of population growth and technological advancements, however, humans have had a great impact on our global environment and climate. Give one person a chain saw and he will do little damage – but, if 6.5 billion people have chain saws the destruction can be catastrophic.

Geologists such as Dr. Jan Zalasiewicz at the University of Leicester, UK are now redefining this interval of geologic time, so heavily influenced by humans, as the "Anthropocene", a word derived from the Greek roots "anthropo-" meaning human, and "-cene" meaning new. The term's origin can be traced back to New York Times journalist Andrew Revkin's 1992 book *Global Warming: Understanding The Forecast* in which he stated – "We are entering an age that might someday be referred to as, say, the *Anthrocene*. After all, it is a geological age of our own making".

The technical term *Anthropocene* was coined by Nobel Prize winning atmospheric chemist Dr. Paul Crutzen at the Max Planck Institute for Chemistry in Germany, and Dr. E.F. Stoermer at the University of Michigan, Ann Arbor in a 2000 article that appeared in the *Global Change Newsletter*. They argued that humans have been the dominant geologic force on the Earth's surface since the industrial revolution and James Watt's invention of the steam engine in 1784. In 2003 and again in 2007 Dr. William F. Ruddiman extended the concept of the Anthropocene back ~ 5,000 to 8,000 years ago with the onset of large scale deforestation (8,000 years ago) and large-scale development of rice paddies (5,000 years ago), both fueled by new technology to feed an ever increasing global human population.

This time extension of the Anthropocene back by thousands of years is consistent with the transformation of our global human population from hunter-gathers to agrarians, and a wave of species extinctions (large mammals and land birds) by hunting and habitat destruction. It is also consistent with Charles Mann's 2005 book -*1491:New Revelations of the Americas Before Columbus* in which he argues strongly that the Americas were much more heavily populated and organized into agrarian societies than previously thought - long before Columbus "discovered" America.

In addition, my own research on sediment cores from the Finger Lakes in central New York State, along with several colleagues, supports Ruddiman's argument that "...the Anthropocene started thousands of years ago..." Our data indicate that Native American populations in the Finger Lakes region were large enough and organized enough to cause considerable land-use change as early as ~ 3,000 years ago. This conclusion supports earlier archaeological work by Dr. John Hart at the New York State Museum who has evidence of Native American cultivation of maize and squash by ~ 2,900 years ago.

One significant sign of the Anthropocene can be found by examining historical data since 1958 – "The Keeling Curve" - on the concentration of atmospheric carbon dioxide, with that found in ice cores from Greenland and Antarctica that go back as far as ~ 1 million years. Between glacial and interglacial periods carbon dioxide concentrations in the atmosphere varied between ~ 180 parts per million (glacial periods) and ~ 280 parts per million (interglacial periods). As of 2009 anthropogenic emissions of carbon dioxide into the atmosphere had increased about the same amount – from ~ 280 to 387 parts per million.

This is of great significance in that the carbon dioxide increase over the past ~ 200 years is occurring at a much more rapid rate, and to a greater extent, than it has in at least the past 1 million years. Today (2009) the concentration of carbon dioxide in the atmosphere is ~ 387 parts per million and increasing at a rate of ~ 1.6 parts per million per year. The United Nations sanctioned Intergovernmental Panel on Climate Change (IPCC), consisting of 1000s of international scientific climate experts, concluded in 2007 that most, if not all, of this recent rise in atmospheric carbon dioxide is of anthropogenic (human) origin resulting from the burning of fossil fuels, cement production and deforestation – all of which release carbon dioxide into the atmosphere on a daily global scale. And, carbon dioxide is a well known greenhouse gas.

Because of technology *Homo sapiens* alter their global environment more than any species on Earth. Further evidence of this has recently been reported by soil scientist Dr. Daniel Richter of Duke University's

School of the Environment and Earth Sciences. He and his colleagues state in a scientific paper published in 2008 by the journal *Soil Science* that – "Humankind has thus far transformed more than half of Earth's soils to suit its needs – agriculture and infrastructure chief among them – and has become the preeminent force changing Earth's surface". Soils are critical to our human global society in that they are the materials from which we grow our plant food and graze our meat animals for food and survival. According to Richter about one-twentieth of soils in the United States are in danger of "substantial loss" or "complete elimination" by a combination of acid rain, pollution and erosion. Dr. David R. Montgomery, a geomorphologist at the University of Washington in Seattle, agrees with Richter in his 2007 book *Dirt: The Erosion of Civilizations*. Montgomery focuses on the interplay and co-dependence of humans and soils for the sustainability of societies and civilizations.

By examining human population and technology from scientific and anthropological perspectives it becomes apparent that our global industrialized society is a product of both an exponentially expanding human population and very rapid technological advancements. Medical technology has interrupted the Earth System's "natural" mechanisms of population control, and technology has provided us with weapons of mass destruction. This raises a philosophical question as to whether or not humans are part of the "natural" system. Many have argued that we are not – thus the concept of the Anthropocene. I, however, am one of many who argue that we evolved within the Earth System over time, and thus we *are* part of that natural system. It is arrogant for us to think of ourselves as anything more than just another species of life on Planet Earth within the Earth System!

V) ARMAGEDDON?

Webster's dictionary defines Armageddon as a 14th century word meaning either "the site or time of a final and conclusive battle between the forces of good versus evil", or "a usually vast, decisive conflict or confrontation". If we greatly exceed the carrying capacity for humans within the Earth System we may well reach Armageddon. We may be in an ultimate battle amongst ourselves, or against the Earth System. If this occurs, some will put their faith in God or technology to save us, whereas others may fear that our species will become extinct like so many of our predecessors over geologic time.

In a November 2007 interview with the father of the Gaia Theory, Dr. James Lovelock, *Rolling Stone*

Magazine refers to Lovelock as "The Prophet". The magazine correctly recognizes Lovelock as "...one of the most eminent scientists of our time..." and suggests that more than 6 billion people will perish by the end of the 21st century. Lovelock believes that climate change is a natural response of the Earth System "...to get rid of an irritating species: us humans, or at least cut them (us) back to size..." And, as my paleontological friend and colleague Dr. Linda Ivany at Syracuse University reminds me, 99.9999999.....% of all the species that have ever lived on Planet Earth since 3.8 billion years ago are now extinct. With that statistic in mind, Lovelock's warning does not seem so unrealistic and at those percentages, the long-term survival of our species can certainly be questioned.

Many leading experts, however, dismiss Lovelock's "doomsday science" claiming that there is no single "tipping point" for humans on a global scale. It is not known, which point of view will prove to be correct, but are we as a species so arrogant as to assume at the 100% confidence level that Lovelock is wrong?

For the purpose of this book, and as an intellectual exercise, let us assume that at some time in this century global human population will significantly exceed its carrying capacity in the Earth System – like the largemouth bass in the farmer's pond. If we are, as Lovelock suggests, a "pest" to the Earth System, what potential options does the Earth System have to exterminate the "pest"?

War: We are a species that wars against our own kind – it seems to be encoded in our genes and DNA. Perhaps similar to Goodall's observation that chimpanzees become more aggressive when crowded together or forced to compete for food. If you are reading this book, you have never experienced an extended period of time when someplace on Earth there was not a war going on. I was born shortly after WW II and have witnessed The Korean War, The Vietnam War, The Persian Gulf War, The War on Afghanistan, The Iraqi War, a seemingly infinite number of Israeli-Arab conflicts, and now the Global War on Terrorism - all in the past 57 years.

There have been thousands of military conflicts over the past 5,000+ years of human history. In the beginning, wars were likely small raids on neighboring tribes, but the invention of gunpowder and numerous technological advancements have resulted in modern global warfare. And, at this moment in time (March 2009) North Korea (a nuclear weapons country) is warning of war against South Korea and the United States (also a nuclear weapons country). The *Associated Press* reported that North Korea has put its troops on military alert as South Korea and the United States prepare for joint military maneuvers. North Korea has warned "...that even the slightest provocation could trigger war". By "provocation" the North is referring to its impending launch of a satellite into orbit that the United States and Japan believe is a cover for a long range ballistic missile test, and they have hinted that the United States may attempt to intercept the rocket launched by North Korea later this month. The North Koreans responded by saying "Shooting our satellite for peaceful purposes will precisely mean a war...that will draw a just military strike..." In response, South Korea has put its military on standby for combat. In reality, much of this may well be "saber rattling" rhetoric, but it points directly to the potential for yet another war – and a nuclear one at that!

War is the "...reciprocal and violent application of force between hostile political entities aimed at bringing about a desired end-state via armed conflict..." The causes of war are multiple and have been studied extensively by historians, sociologists, psychologists, political scientists and economists for many years. According to Dr. R.J. Rummel Professor Emeritus of Political Science at the University of Hawaii "...war arises because of the changing relations of numerous variables – technological, psychic, social, and intellectual…" It is aggravated by socio-cultural dissimilarity, cognitive imbalance, status difference, or coercive power. It is inhibited by socio-cultural similarity or decentralized state power. It is triggered by perception of opportunity, threat, injustice, or surprise.

There are numerous theories as to why humans go to war with the claim of moral justification. The "tradeoff analysis theory" is based on the perception that the benefits of war will be greater than the cost. Justification may be to protect national pride, preventing loss of territory, or to inflict punishment. "Behavioral theory" argues that humans are inherently violent and wars are waged based on bias and hatred against other ethnic or racial groups, different religions, or ideologies. Peace never really exists and the hope of escaping war is limited. War is a part of animal behavior to protect territory and to compete. "Sociological theory" suggests that WW I was, in large part, the product of economic, social, and political differences.

"Demographic theories" are grouped into two classes. Malthusian theory is based on an ever expanding human population and limited resources that ultimately leads to violent conflict. Of all the theories discussed, Malthusian theory is most closely aligned with Earth System Science concepts. Thomas Malthus (1766-1834) wrote that "...populations always increase until they are limited by war, disease or famine..."

The "youth bulge theory" is based on a disparity between "excess" young males of fighting age (15-29 years) and limited availability of social or economic status. Religion and ideology are seen as secondary factors. Dr. Samuel Huntington (1927-2008), a former professor of government at Harvard University and later Columbia University, wrote – "I don't think Islam is any more violent than any other religion, and I suspect if you added it all up, more people have been slaughtered by Christians over the centuries than by Muslims....during the 1960s, 70s and 80s there were high birth rates in the Muslim world.... (that gave) rise to a huge youth bulge". War is, thus, a result of "many angry young men" finding themselves marginalized in a global society. Both Christian European colonialism/imperialism and today's Islamic unrest and terrorism are both thought to be, in large part, a consequence of high birth rates and youth bulges. The French Revolution of 1789, the rise of German Nazism in the 1930s, and the 1994 Rwandan genocide are also thought to have been a product of a "youth bulge" – human overpopulation.

"Rationalist theory" assumes that both parties to a conflict are rational and try to achieve the best possible outcome for themselves with the least loss of life and property. The United States entered the Vietnam War with full knowledge of resistance by the North Vietnamese, but misinformed about their long-term capability of guerilla warfare. The "economic theory" sees war as a natural "outgrowth" of economic competition on an international basis. Fascists assert a "natural right" of the strong to conquer the weak. Even U.S. President Woodrow Wilson stated in 1919 that "…is there any man, is there any woman, let me say child here that does not know that the seed of war in the modern world is industrial and commercial rivalry?"

"Marxist theory" believes that all wars are waged based on class struggle – the so-called "haves" and "have-nots" – in order to divide the world into a "ruling class" and a sub-servant "proletariat". Marxists argue that war is a result of the free market and class system and will not stop until there is a global revolution. "Political theory" states that the motivation for war is security (survival) that goes in cycles in which one nation wants to dominate the world. WW II, the largest war in human history with ~ 71.1 million people killed, is commonly viewed as a transitional political war.

In order to conduct war nations need to have weapons, technology, armed forces, tactics, strategy, and logistics which are all part of military science. And, there are some aspects of war that some people considered positive, such as the reduction of population and economic stimulus.

Nuclear War/Winter: Toward the end of WW II the United States, with assistance from the United Kingdom and Canada, formed the Manhattan Project to develop a nuclear weapon under the research direction of physicist Dr. J. Robert Oppenheimer. By order of U.S. President Harry S. Truman, the face of war changed dramatically forever on August 6, 1945 when the B-29 aircraft *Enola Gay*, piloted by Colonel Paul Tibbets dropped the first atomic bomb ("Little Boy") on the Japanese city of Hiroshima killing an estimated 140,000 people. Three days later, the B-29 aircraft *Bockscar*, piloted by Major Charles W. Sweney, dropped a second atomic weapon ("Fat Man") on the city of Nagasaki killing an estimated 80,000 people, prompting the unconditional surrender of Japan and the end to WW II. A mere 64 years ago we entered the nuclear age of global warfare!

Atomic bombs (A-bombs) use fissile (ability to be split) material such as enriched uranium or plutonium at supercritical mass to produce an exponentially

43

growing nuclear reaction with the energy of up to ~ 500,000 tons of TNT. Hydrogen bombs (H-bombs) are more powerful as they rely on the fusion reaction of two isotopes of hydrogen to create a nuclear reaction that has no inherent limit on energy released. The first H-bomb was tested by the United States in 1952 on Enewetak Atoll (a coral island) in the southwest Pacific Ocean. Today, only five countries (United States, Russia, Great Britain, France and China) have

tested H-bombs and none have yet to be used in warfare. The largest test was conducted by the former Soviet Union that had an equivalent of 50,000,000 tons of TNT. Today countries that are acknowledged to have nuclear weapon capabilities are Britain, China, France, India, Pakistan, Russia, United States, and North Korea. An unacknowledged country, Israel, likely has nuclear weapon capability. Iran is a country that is apparently seeking to have nuclear weapons. In a March 2, 2009 news report, Admiral Mike Mullen, chairman of the Joint Chiefs of Staff for the United States was asked if Iran has enough fissile material to make a nuclear weapon and he responded that "We think they do, quite frankly. And Iran having a nuclear weapon…is a very, very bad outcome for the region and for the world." However, U.S. Defense Secretary Robert Gates, in the same report, stated that the Iranians are "…not close to a weapon at this point…" in time. A number of countries have actually abandoned their nuclear weapons including South Africa, Belarus, Kazakhstan, and the Ukraine. However, there are a total of 40 countries with enough plutonium and/or uranium to produce nuclear arsenals, and interest in using nuclear energy to

combat global warming is giving many other countries the technology to potentially "go nuclear".

As a result of the SALT 1 Treaty (1969-1972) stockpiles of nuclear weapons have been on the decline from ~ 65,000 warheads in 1985 to ~ 20,000 as of 2002. The United States and Russia have the most nuclear warheads (~ 4,000 to up to 9,000 each), and North Korea the least with less than 10. In the decade following 2007 there is an expected additional decline of 30% to 50%, although 1000s of nuclear warheads have only been dismantled not destroyed. The exact number of A-bombs versus H-bombs in these arsenals is not well known.

So is there the *potential* for global nuclear war? The answer is an unequivocal yes. Iran has openly declared its desire to destroy Israel. If Iran attacks Israel with nuclear weapons, will not Israel respond in kind? If the United States are attacked by terrorists with nuclear weapons will it not respond in kind? If North Korea attacks South Korea or Japan with nuclear weapons will not the United States respond in kind?

I am old enough to vividly remember the Cuban missile crisis in the early 1960s between the United States, the Soviet Union and Cuba. It is considered the closest moment during the Cold War that we came to nuclear war. The crisis started on October 8, 1962, and on October 14, 1962 United States spy planes confirmed that offensive missile bases were being built in Cuba by the Soviets. The crisis ended on October 28, 1962 when the President of the United States and the United Nations Secretary General made an agreement with the Soviets to dismantle their missile bases in Cuba in exchange for a "no invasion" agreement and removal of offensive nuclear missiles in Turkey by the United States. President John F. Kennedy put the world on notice in a speech he gave on October 22, 1962 – "It shall be the policy of this nation to regard any nuclear missile launched from Cuba against any nation in the Western Hemisphere as an attack on the United States, requiring full retaliatory response upon the Soviet Union". The world was on the precipice of global nuclear war!

Since the Cuban Missile Crisis, global diplomacy and agreements have reduced nuclear arsenals and have thus far kept the world from nuclear war. Also since 1962, military technology has increased dramatically, human population has continued to rise rapidly, there are new nuclear countries, and there are new threats to "national security". Thus, the *potential* for global nuclear war still exists, as does its potential consequence – "nuclear winter" – which has the *potential* to result in the extinction of *Homo sapiens.*

Nuclear winter is "…a worldwide darkening and cooling of the atmosphere with consequent devastation of surviving life forms, believed by some scientists to be a probable outcome of large-scale nuclear war…" In 1982 Dr. Paul Crutzen, then at the University of Colorado and Dr. John Birks at Germany's Max Planck Institute suggested that large-scale nuclear war would result in global wildfires that would put enough smoke and soot into the atmosphere that would block sunlight, inhibit photosynthesis, cause global cooling and catastrophic agricultural failure.

Their suggestion was quickly analyzed with quantitative computer models by Dr. R.P. Turco at R

& D Associates in Marina Del Ray, California, the late astrophysicist Dr. Carl Sagan at Cornell University and their colleagues who concluded that large-scale nuclear war would result, from the burning of cities and forests, in a global temperature drop of 36° C (97° F) that would last for several months. Such a temperature drop is enormous when one considers the fact that today's mean annual global surface temperature on Earth is only 15° C (59° F). They further concluded that this temperature drop would result "...in the greatest biological and physical disruption of the planet in its last 65 million years...and that the number of survivors would be reduced to prehistoric levels...a fraction of 1 % of those now alive..." Their computer models became known as the TTAPS study, and Turco and Sagan coined the term "nuclear winter" to describe the phenomenon.

There were a number of uncertainties in the TTAPS study, and by 1989 Turco and Sagan had developed five different scenarios of nuclear winter that had as much to do about global political policy as geophysics. By the early 1990s, nuclear winter scenarios fell out of favor due to the above mentioned uncertainties, because the Kuwait oil fires from the Persian Gulf War did not lead to substantial global climate effects (only regional cooling of 4-6° C = 4° F), and concern about nuclear war had waned because of the end of the Cold War. There are, however, two common misperceptions about nuclear winter: 1) that the theory has been scientifically disproven; and, 2) even if correct, the theory is irrelevant because of the end of the nuclear arms race. These are very dangerous assumptions to make!

Global fires in both cities and forests would release smoke and aerosol (very small droplets) through the Earth's troposphere in which we live, and up into the stratosphere 10-15 km (6-9 miles) above the Earth. The smoke and aerosols would reach the stratosphere by heating from the sun that would cause it to become less dense and rise. This is an extremely important point because weather occurs only in the troposphere where atmospheric precipitation would quickly remove the "pollutants". But, in the stratosphere there is essentially no weather that would allow the pollutants to remain in the upper atmosphere for many years to block sunlight, limit photosynthesis and cause a large-scale decrease in global temperature. With this knowledge in hand, scientists have conducted several recent studies of the nuclear winter hypothesis.

In 2006 a team of scientists from Rutgers University, the University of Colorado at Boulder, and the University of California at Los Angeles reported twin papers at the national *American Geophysical Union Conference* in San Francisco that a regional nuclear war in which the combatants both used 50 Hiroshima sized A-bombs would result in human fatalities ranging from 2.6 to 16.7 million people in both of the combatant countries. The explosions would also release ~ 5 million tons of soot that would result in several degrees of cooling over large areas of North America and Eurasia, including most grain-growing regions. The cooling would last for years with potentially catastrophic results.

In 2008 a study published in the *Proceedings of the National Academy of Science* examined a nuclear exchange between Pakistan and India using their current arsenals of nuclear weapons. Their conclusion was that this would create a near-global ozone hole in the stratosphere that would wreak environmental havoc for decades. Ozone is of critical importance to life on Earth as it protects us, and other organisms, from high-energy ultraviolet radiation (UV) emitted by the Sun. The computer model assumed that both India and Pakistan each fired 50 Hiroshima sized A-bombs that would send a cloud of pollution ~ 80 km (50 miles) into the stratosphere which is the portion of the atmosphere where ozone is produced and where the ozone layer resides.

Research results published by some of the pioneers of nuclear winter in 2007 and 2008 confirm that it is still a very viable hypothesis for mass extinction of humans, despite the fact that the present nuclear weapons treaty between the United States and Russia requires both countries to have only about 2000 nuclear warheads by 2012. But remember that many of both countries' nuclear weapons have only been dismantled not destroyed. Even with the explosion of ~ 4000 A-bombs, the result would be the release of ~ 150 megatons (millions of tons) of smoke and aerosols that would lead to unprecedented rates and range of global temperature decreases. This would result in true nuclear winter making agriculture impossible for years with effects lasting for decades.

Dr. Alan Robock and his colleagues in the Department of Environmental Sciences at Rutgers University and in the Department of Earth and Planetary Science at Johns Hopkins University conducted a new study of nuclear winter in 2007 using a computer model and concluded that there are "…still catastrophic consequences…" to nuclear winter. In their model they assumed an input of 150 megatons of smoke as a "worst case scenario" for nuclear weapons in arsenals held by the United States and Russia in 2012. Their results suggest "…a global average surface cooling of -7 to - 8° C (45-46° F) persisting for years, and after a decade the cooling is still -4°C (39° F). Considering that the global average cooling at the depth of the last ice age 18,000 years ago was about 5° C (9° F), this would be a climate change unprecedented in speed and amplitude in the history of the human race. The temperature changes are largest over land...cooling of more than -20° C (-4° F) occurs over large areas of North America and more than -30° C (-22° F) over much of Eurasia, including all agricultural regions". They also discovered a possible weakening of the global hydrologic cycle that would result in a decrease of global precipitation by ~ 45%, and that the lack of food production would last for many years, "...making the impacts of nuclear winter even worse than previously thought..."

Nuclear winter scientists are not without their critics. The original TTAPS studies of Turco and Sagan were widely reported and critiqued by the media. And, in 1987 Dr. Cresson Kearny of the Oak Ridge National Laboratory referred to the TTAPS studies as a "propaganda story", a "myth", and a "discredited theory". And in 1988 Dr. Brian Martin in the School of Social Sciences, Media and Communications at the University of Wollongong, Australia preferred the term "nuclear autumn" rather than nuclear winter because it was his belief that the effects of nuclear war would be less than that originally proposed by Turco, Sagan and their colleagues. Also, in 1986 Russell Seitz reported in *The National Interest* that the prominent British-born, American physicist and mathematician Dr. Freeman Dyson, who taught at both Cornell University and Duke University, stated that the TTAPS study was "...an absolutely atrocious piece of science...but who wants to be accused of being in favor of nuclear war?". Seitz also stated in the scientific journal *Nature* that nuclear winter research "...has become notorious for its lack of scientific integrity..." However, all these critical comments were made in the late 1980s soon after the nuclear winter hypothesis was first introduced, and before it has more recently been confirmed. Thus, nuclear winter continues to be a viable scenario as an extinction mechanism for *Homo sapiens*!

Biological Warfare: Biological warfare, or more specifically bio-terrorism, is another concern for large-scale mass destruction of human life. Biological warfare is the use of pathogens (bacteria, viruses, or other disease-causing agents) as biological weapons intended to kill, incapacitate, or seriously impede entire cities or places as a military tactic. Similarly, bio-terrorism is the intentional release or dissemination of biological agents (bacteria, viruses, toxins) to cause illness or death of people, animals or plants. It is used by terrorist groups mostly as a method of creating mass panic and disruption to a given society. But with our increasing technology in bioengineering, bioterrorism may become a very realistic threat to our global population.

The Center for Disease Control categorizes biological agents into three groups. "Category A" agents have both a high potential for impacting public health and a serious potential for large dissemination by air, water or in food. Category A agents include anthrax which is non-contagious, smallpox that is a highly contagious virus, botulism which is one of the deadliest bacterial toxins, the plague which is bacterial in origin, hosted in rodents and transmitted to people via by flea bites, and viral hemorrhagic fever that kills by multiple organ failure and shock. "Category B" agents are numerous, but only moderately easy to disseminate with low mortality rates. And, "Category C" agents are pathogens (a causative agent of disease) that may be engineered for mass dissemination because they are easy to produce and potentially have very high mortality rates.

The creation and stockpiling of biological offensive weapons was outlawed in 1972 by the international Biological Weapons Convention with the rationale of avoiding a successful biological weapon attack that could conceivably kill millions, if not billions, of people. According to the U.S. Office of Technology Assessment, although outlawed, numerous countries were believed to posses biological weapons in 1995, including Libya, North Korea, South Korea, Iraq, Taiwan, Syria, Israel, Iran, China, Egypt, Vietnam, Laos, Cuba, Bulgaria, India, South Africa and Russia as well as the United States.

Biological warfare has been occurring throughout much of human history. Before the 20th century the use of biological agents was in the form of deliberate poisoning of food, use of microorganisms, toxins or animals as weapons, and/or the use of biologically infected fabrics. The first documented case of biological warfare occurred between ~ 1500-1200 BC by driving victims of plague into enemy lands. During the Trojan War, spears and arrows were tipped with poison, and in 590 BC the Greeks intentionally poisoned water supplies of their enemies.

The Monguls catapulted the corpses of plague victims over the walls of enemy cities. In 184 BC, Hannibal ordered his troops to throw pots filled with venomous snakes into the enemy, and in 198 Iraqis turned back the Romans by throwing pots filled with live scorpions. The last known incident of using plague corpses was in 1710 when Russians attacked the Swedes. In the same year, Native Americans used biological weapons against the British by poisoning water sources with dead animals that infected the water with *E. coli* bacteria. Native American populations were also devastated by disease by coming into contact with Europeans either by intent (blankets infected with smallpox) or lack of immunity to Old World diseases.

During WW I, Germany had a very active biological warfare program, and sent small groups of saboteurs equipped with biological weapons into enemy countries. The Geneva Protocol of 1925 banned the use of biological weapons, but allowed for their production, storage and transfer. Technological advances allowed for the first pure-culture biological agents to be developed by a number of countries, and Japan used biological weapons for attacks in China during WW II.

In 1940 Japan dropped bombs filled with fleas containing the bubonic plague in China. In response to the German and Japanese use of biological weapons, the United States, the United Kingdom and Canada initiated a biological weapons program in 1941 at Fort Detrick, Maryland. In 1948, the Red Cross accused the Jewish Army of releasing *Salmonella* bacteria into enemy water supplies. During the Cold War China, North Korea and Cuba accused the United States of using disease-carrying insects against them. In 1953 the United States developed biological weapons such as plague-fleas, EEE-mosquitoes and yellow fever-mosquitoes that could be dispersed by "cluster bombs" or in the warheads of missiles. Dry-based biological agents, resembling talcum powder, were also developed that could be disseminated as aerosols via gas explosions.

In 1969 President Richard M. Nixon signed an executive order to stop the production of biological weapons in the United States, but allowed their use in scientific experiments, as defensive agents and in bio-safety. The Central Intelligence Agency (CIA), however, had covert biological weapons as late as 1975. In 1972 the United States signed the Biological and Toxic Weapons Convention that banned "the development, production and stockpiling of microbes or their poisonous products except in amounts necessary for protective and peaceful research".

By 1996, 137 countries had signed this Convention, but since then the number of countries capable of producing biological weapons has actually increased. After the 1991 Persian Gulf War, Iraq admitted to the United Nations that it had produced ~ 19,000 liters (5020 US gallons) of concentrated botulism toxin of which about half were loaded into military weapons. This total volume of toxin is about 3 times the amount needed to kill the entire human population on Earth via inhalation! Unfortunately, the 19,000 liters have never been fully accounted for.

The ideal characteristics of biological weapons used to attack humans are that they be highly infectious, strong, immune to vaccines, and be delivered as aerosols that can be widely dispersed. They also need to be made quickly, easily and deliverable to the intended target. For example, pneumonic anthrax infection has a 90+ % fatality rate, but can be protected by antibiotics – thus it would not be an "ideal" biological weapon.

There are, however, numerous biological agents that do fall into the "ideal" category. And some of these agents can also be used to destroy food crops on a large scale. During the Cold War the United States developed anti-crop capability to thwart aggression by the Soviet Union, and the Soviet Ministry of Agriculture developed biological agents to target cows, pigs and chickens to eliminate enemy food sources, as well as agents to be delivered to enemy watersheds to initiate epidemics of plants. Ironically, this Soviet military venture was code-named "Ecology".

Today, western countries like the United States are now focused, not on the production and use of biological weapons, but the early detection of such agents (bio-defense) because biological agents are relatively easily obtained by terrorist groups and thus pose an increasing threat. The basic concept is to have an early warning system to allow for rapid treatment. In 1999, the University of Pittsburgh's Center for Biomedical Information developed the first automated bioterrorism detection device referred to as "RODS" – Real Time Outbreak Disease Surveillance.

Chemical Warfare: Chemical warfare by definition is the use of toxic properties of chemical substances as weapons to kill, injure or incapacitate an enemy. However, the use of nonliving toxins produced by organisms is also considered chemical warfare. Approximately 70 different chemical substances have been used or stockpiled during the 20th century, and according to the United Nations, these chemicals are "weapons of mass destruction".

Their use and stockpiling were outlawed by the United Nations' Chemical Weapons Convention (CWC) in 1993. The UN classifies chemical weapons into three types: 1) Schedule 1 – those that have limited or no legitimate use such as nerve agents, ricin (poisonous protein in castor beans), or mustard gas; 2) Schedule 2 – those that may have legitimate small scale use such as sarin (a nerve gas) used as a flame retardant; and, 3) Schedule 3 – those that have legitimate large scale industrial use in plastics and fumigants.

Rudimentary chemical warfare has been conducted for thousands of years. Leonardo da Vinci, during the 15th century has been quoted as writing - "...throw poison in the form of powder upon galleys. Chalk, fine sulfide or arsenic and powdered verdigris (poisonous copper powder) may be thrown among enemy ships by means of small mangonels (military missiles), and all those who, as they breathe, inhale the powder into their lungs and will become asphyxiated".

Modern chemical warfare began with WW I. Germany was the first country (1917) to use modern chemical warfare by opening cans of chlorine upwind of the enemy. Also during WW I, France developed artillery shells filled with phosgene that inflicts its victims with burns and blisters. Since WW I numerous countries have developed and stored more deadly chemical agents and improved on their methods of dispersal. For example, new liquid chemical weapons are intentionally made to be "volatile" (high vapor pressure) so that they may be disseminated rapidly over large areas. Non-volatile liquid chemical weapons, such as blister and some nerve agents, require direct contact to kill an enemy. Chemical weapons not covered by the 1993 UN

Convention include Agent Orange (a defoliant) and napalm (an incendiary chemical) used by the United States during the Vietnam War.

The nations of the North Atlantic Treaty Organization recognize six different types of chemical agents: 1) Blood agents - such as cyanogens (colorless, flammable, poisonous gas) that are designed to attack blood cells and inactivate their energy producing enzymes; 2) Pulmonary agents – such as phosgene and chlorine that induce choking, limit breathing and cause death by suffocation; 3) Lachrymatory agents – that cause severe blisters via chemicals such as bromides and chlorides; 4) Vesicant agents – that cause chemical burns using sulfur-mustard or nitrogen-mustard; 5) Incapacitating agents – designed to cause temporary physiological or mental effects but can be lethal; and, 6) Nerve agents – made from organophosphates that disrupt neurotransmission in the body and cause death by asphyxiation due to loss of respiratory muscles.

The most important component in chemical warfare is not the agents discussed above but the dissemination or delivery of those agents to the chosen target. Common techniques include bombs, missile warheads, and spray tanks from airplanes. Dissemination is complicated by weather conditions because many agents are delivered in a gaseous form.

During the 1950s and 1960s delivery technology improved with the development of artillery rockets and cluster bombs, as well as thermal dissemination using explosive or pyrotechnic devices.

Chemical warfare was a common practice in China and India, but has been viewed as "immoral" in much of the western world. Despite this, research and development in both the United States and Russia led to the proliferation of chemical weapons during the Cold War. Former U.S. Vice President Dick Cheney opposed the U.S. signing of treaties to ban the use of chemical weapons in 1997 because "Those nations most likely to comply with the Chemical Weapons Convention are not likely to ever constitute a military threat to the United States. The governments we should be concerned about are likely to cheat on the CWC, even if they do participate". Although the CWC was ratified in 1997 Albania, Libya, Russia, India and the United States have declared more than 71,000 metric tons (78,100 tons or 156.2 million pounds) of which only about one third has been destroyed. Russia and the United States are "required" to destroy their remaining stockpiles between 2012 and 2017.

During WW II the Imperial Japanese army used chemical weapons extensively, particularly against other Asian nations judged to be "inferior". Mustard gas and blister agents were used profusely against the Chinese, and for defensive purposes Australia imported ~ 1 million chemical weapons from the United Kingdom in 1942. On December 2, 1943 Germany attacked the port of Bari in Southern Italy and sank numerous American ships carrying mustard gas – an attacked dubbed "The Little Pearl Harbor". And, Islamic and Palestinian leaders aligned with Adolf Hitler were accused of an unsuccessful chemical attack on Tel-Aviv in 1944.

During the Cold War there was extensive research, development and stockpiling of nerve agents by all of the Allies as well as the Soviet Union. In 1958 the British government sent VX nerve agent technology to the United States in return for technology on nuclear weapons. By 1961 the United States was producing vast quantities of nerve agents collectively known as "V-Series" weapons. On April 4, 1984, U.S. President Ronald Reagan proposed an international ban on the use of chemical weapons, which was subsequently signed as a bilateral treaty by U.S. President George H.W. Bush and Soviet leader Mikhail Gorbachev in 1993 and put into effect in 1997.

The most serious use of chemical weapons occurred during the Iran-Iraq War by Saddam Hussein who acquired the technology for chemical warfare from the United States, West Germany, the Netherlands, the United Kingdom, France and China. During this war, which began in 1980, Iraq delivered mustard gas via bombs from airplanes that resulted in an estimated 5 % of all Iranian casualties. Approximately 100,000 Iranian soldiers were killed by mustard gas and 20,000 by nerve agents. Near the end of the war Iraq used multiple chemical weapons on Kurdish people in its own country killing at least 5,000 people.

For terrorist organizations, chemical weapons are ideal because they are cheap, accessible, and easy to transport. The first successful terrorist attack occurred in Japan in 1995 when a terrorist group released sarin into Tokyo subways killing 12 people and injuring more than 5,000. And, after the September 11, 2001 Al Qaeda attack on the World Trade Centers in New York City and other U.S. targets, the terror group stated that they were acquiring large stockpiles of not only chemical weapons, but also biological and radiological weapons that was verified by the television news station CNN in 2002. In 2007 numerous terrorist chemical weapon attacks occurred in Iraq using chlorine gas delivered by tanker trucks. Like biological weapons, there are multiple chemical weapons readily available to any hostile nation or terrorist group.

Pandemics: By definition, a "pandemic" (derived from the Greek word *pan* meaning all and *demos* meaning people) is an epidemic of infectious disease that spreads through populations across a large region such as a continent or even the globe. The World Health Organization (WHO) requires three criteria to recognize a pandemic: 1) emergence of a new disease to a large population; 2) the agent infects humans causing serious illness; and, 3) the agent spreads easily and sustainably among people. It *must* be infectious. Cancer, for example, is not considered a pandemic because it is not infectious.

Pandemics appear to be a preferred manor of human population control by the Earth System. It is a negative feedback that can potentially keep human population below the Earth System's carrying capacity for it. Examples include the Peloponnesian War in 430 BC when bacterial typhoid fever killed ~ 25 % of the population of Athens in four years. The Antonine Plague (165 to 180) was caused by smallpox that invaded the Italian Peninsula and killed up to 5 million people. At is peak ~ 5,000 people per day were dying in Rome. Between 541 and 750, the first recorded outbreak of bubonic plague started in Egypt and spread to Turkey killing up to 10,000 people per day. It resulted in a ~ 50% reduction in Europe's human population. The "Black Death" (also bubonic plague) between 1347 and 1352 killed ~ 25 million Europeans or about 33% of population at that time. Another example is the American pandemic caused by a host of pathogens that killed ~ 50 million Native Americans between 1492 and 1700 when Native peoples with low immunity to these diseases came into contact with European settlers – a form of "bio-transport". This American pandemic wiped out ~ 85-90 % of the Native American population at the time.

There have been numerous pandemics during human history, many associated with "zoonoses" resulting from the domestication of livestock. There have been seven cholera pandemics around the world between 1816 and 1966 which spread rapidly across single continents and between continents killing 10s of millions of people. The first cholera pandemic (1816-1826) was one of the worst, killing ~ 100,000 British troops in India before spreading to China, Indonesia and the Caspian Sea. More than 15 million people in India died, ~ 2 million in Russia, and ~ 100,000 on the Island of Java. The second cholera pandemic (1829-1851) started in Russia ("Cholera Riots") and spread throughout the Northern Hemisphere reaching both coasts of the United States by 1834.

Influenza pandemics began in 1510 in Africa and quickly spread across Europe. The "Asiatic Flu" of 1889-1890 began in Europe and reached North

America, South America and Australia with "a very high attack and mortality rate". The "Spanish Flu" of 1918-1919 began at the U.S. military training Camp Funston, Kansas, spread world-wide and killed 10s of millions of people across the globe. The Asian flu of 1957, the Hong Kong Flu of 1968, the Swine Flu of 1976, the Russian Flu of 1977, the Avian Flu of 1997, and the Swine Flu of 2009 all serve as proof that some of the viruses that cause global influenza and pandemics are still in circulation today.

Typhus pandemics are sometimes referred to as "camp fever" or "ship fever" due to its occurrence during times of strife or in close quarters, such as on a ship or in jails. During the Thirty Years' War (1618-1648) about 8 million Germans were killed by a combination of typhus fever and the bubonic plague. Typhus also played a key role in the defeat of Napoleon's *Grande Armee* by the Russians in 1812. Typhus also played a significant role in the Irish Potato Famine of 1845-1852, and was responsible for the deaths of numerous captives in Nazi concentration camps and Soviet prisons during WW II.

"Globalization" was a key contributor to many of these early pandemics as colonial or imperialistic Europeans came into contact with native peoples with low immunity to these diseases. The entire native population of the Canary Islands was wiped out in the

16th century. Old World diseases were responsible for the death of up to ~ 95 % of the Native American population after 1491. And about 50 % of Native Australian people were killed by smallpox introduced by colonial Britain. But disease also spread from the New World to the Old World – an example being syphilis that killed many Europeans during the Renaissance Era of 1450-1600.

We are presently in the midst of the HIV-AIDS pandemic that has an infection rate up to 25 % in parts of Africa where the death toll is estimated to reach 90-100 million people by 2025. By definition, acquired immune deficiency syndrome (AIDS) is "a set of symptoms and infections resulting from the damage to the human immune system caused by the human immunodeficiency virus (HIV)". This disease progressively lessens the immune system of its victim and opens it up to "opportunistic" infections and tumors. HIV is transmitted by contact of bodily fluids associated with sex, IV-drug use, blood transfusions and breast feeding.

AIDS is considered a pandemic. As of 2007 approximately 33 million people were infected with the disease that has already killed more than 2 million people worldwide. More than 75 % of these deaths have occurred in sub-Saharan Africa where HIV is believed to have originated during the 20th century. It was first recognized by American and French scientists in 1981. There is presently no vaccine or cure for AIDS, so the focus has been placed on prevention. "Antiretroviral" drugs reduce the aggressiveness of the disease, but these drugs are very expensive, and not readily available to many of our global population.

The opportunistic infections that attack HIV patients include bacteria, viruses, fungi and parasites that the human body normally can combat. Pulmonary infections are common in HIV infected people, as are gastrointestinal infections, neurological infections such as encephalitis and meningitis, and there is a dramatic increase in cancerous tumors and malignancies in HIV patients. The HIV virus is capable of infecting all organs of the body by directly and indirectly destroying individual cells. The average time of progression from HIV to AIDS is 9-10 years, and the median time of survival once a victim has AIDS is ~ 9-10 months. Not only do we presently live in the HIV-AIDS pandemic, there is considerable concern about future pandemics. Lassa fever, Rift Valley fever, the Marburg virus, the Ebola virus and Bolivian hemorrhagic fever are very contagious diseases with very high mortality rates that have potential to blossom into pandemics.

Of particular concern is that some of these pathogens are becoming resistant to antibiotic medical treatment – "superbugs". Tuberculosis and smallpox, once thought to be "well-controlled", are making a comeback. Drug resistant tuberculosis was discovered in Africa in 2006 and has since spread to at least 17 countries including the United States. Over the past 20 years, very common bacterial infections such as *Staph* have developed resistance to some of our most potent antibiotics.

As of 2008 "...the threat of death-defying bacteria...is growing more alarming". Dr. Stuart Levy, a microbiologist at Tufts University in Boston, stated that "...this is very worrisome...we are not keeping up with the bacteria". According to Dr. Levy, there are two major problems. First, some bacteria have developed the ability to literally "eat" the antibiotics sent to "eat" them. Dr. George Church, a geneticist at Harvard Medical School, said that these "superbugs" are eating our antibiotics "for breakfast". And secondly, a new and very lethal form of tuberculosis, that is virtually impossible to cure, has recently expanded rapidly in Africa, Asia, and Russia as well as a few cases in the United States. The World Health Organization (WHO) estimated in 2006 that ~ 116,000 deaths occur from this new form of tuberculosis with more than 50 % in China, India and

70

Russia. Dr. Mario Raviglione of the WHO stated that "...if we don't act now, we are really going to risk a disaster of an enormous proportion..." and added that "...this is no joke..."

A final point is that "pandemics" are not restricted to humans – they can wipe out agricultural crops as well. There have been numerous reports of locust swarms destroying crops from biblical times to today. And recently scientists have discovered why locust, that are normally solitary, easy-going creatures swarm. The answer is overcrowding at times of drought and limited food resources that leads to the production of the chemical serotonin, which also controls mood in the human brain. Dr. Malcolm Brown of the University of Cambridge, UK stated "...here we have a solitary and lonely creature, the desert locust. But just give them a little serotonin, and they go and join a gang..." In a paper published in *Science* the researchers discovered that locust triple the amount of serotonin in their bodies when they swarm.

Locusts impact about 20% of the Earth's surface, and last year a swarm about four miles long devastated parts of Australia. Locusts, however, are also found in Africa, Asia and the United States. Population densities of locust swarms may reach into the millions or even billions. This "Dr. Jekyll to Mr.

71

Hide" transformation is stimulated by drought and limited food resources, as well as sight, smell and touch of other locusts – overcrowding.

New agricultural pests are being discovered each and every year. For example, Dr. Susan Halbert, a taxonomic entomologist (study of insects) at the Florida Department of Agriculture and Dr. David J. Voegtlin at the Illinois Natural History Survey have recently (February 2009) reported that the Asian citrus psyllid may be "...the most serious citrus pest in the world..." and that it is now established in Florida. The pest damages crops by transferring pathogens to citrus trees causing "yellow dragon disease" in Chinese, or "yellow shoot disease" in English. Curiously, these pests are very small, about the size of a pinhead (0.16 inches), but can cause great devastation to citrus crops. This Asian citrus pest is also heading north from Mexico into California that has a billion dollar citrus industry. Gary Moles of Willits and Newcomb Nursery stated that "The disease is death to citrus trees. There's no cure for it." The pest kills citrus trees by stopping the flow of fluids and nutrients to the tree, resulting in color change of the fruit from orange to green or yellow. It has already created major devastation to citrus growth in Florida, Texas and Louisiana, and is now in San Diego County, California.

Thus, pandemics, whether they affect humans directly or indirectly by reducing our food supply, are a serious threat to world population especially considering the development of "superbug" bacteria that are resistant to our most potent antibiotics.

Famine: Do we presently have a global food crisis? The answer is an "unequivocal" yes and no. If you live in an industrialized, western country you probably do not even think of the possibility of famine. But, if you live in Haiti, Cameroon, Kenya, Cambodia or other developing countries, the answer is yes. In Haiti where ~ 75 % of the population earns less than $2 per day, many children literally eat "mud patties" that contain vegetable oil, salt and sugar that "...makes your stomach calm quite down..." so that "...you don't know you are eating dirt..."

But if there is truly no threat to biological diversity and of global famine, why do we have the "Ark of the Arctic" and human fertility banks? The "ark" is a global seed vault dug into a mountainside of Norway's Svalbard islands. It will contain some 4.5 million agricultural seeds as a "safety net" for our global crop diversity. It is considered by many as a "doomsday vault" in response to worries about global catastrophe. The vault will protect agricultural seeds from major disasters such as war or global warming.

73

The United States has its own seed bank in Fort Collins, Colorado. The "Ark of the Arctic" was a direct result of the 2001 International Treaty on Plant Genetic Resources and the United Nations' Millennium Development Goals to maintain agricultural diversity and eliminate hunger.

Hundreds of famines have inflicted great distress on human population over history. Famine by definition is "a widespread shortage of food... usually accompanied by regional malnutrition, starvation, epidemics and increased mortality...typically linked to overpopulation, as the number of humans exceeds regional carrying capacity". The poor are very susceptible to famine because of their inability to command sufficient food resources for survival – the "entitlement theory" of Dr. Amartya Sen, Professor of Economics at Harvard University.

Several famines have also been caused by a combination of political, economic, biological and/or climatologic factors, and exacerbated by poor governance and infrastructure. The Great Famine in Ireland that began in 1845 was largely a product of food being shipped from Ireland to England, and the Chinese famine of 1958-1961 was a result of Mao's "Great Leap Forward" communist philosophy, as was North Korea's famine in the 1990s. Climate change, producing unprecedented droughts in regions of

74

Africa and land-use changes can also cause famines. About 40 % of the world's agricultural land is highly degraded, and because of this it is estimated that by the year 2025, Africa will only be able to sustain ~ 25 % of its growing population.

Technological advancements in agriculture such as nitrogen fertilizers, pesticides, desert farming, and farm machinery increased world grain production by some 250 % between 1950 and 1984, the so-called "Green Revolution". However, this enormous increase in food production is not sustainable due to the decline of arable lands and soil degradation. Dr. David Pimentel, Professor of Ecology and Agriculture at Cornell University and Dr. Mario Giampietro at the National Research Institute on Food and Nutrition, in their 1994 report – *Food, Land, Population and the U.S. Economy* - estimate that the sustainable agricultural carrying capacity for the United States is about 200 million people. If correct, the United States will have to reduce its population by at least one-third, and the world population will have to be reduced by about two-thirds! These authors argue that we will not begin to feel the impact of this until 2020 and that it will not become dangerous until about 2050. Michigan geologist Dr. Dale A. Pfeiffer has also concluded that "...the coming decades could see spiraling food prices without relief and massive starvation on a global level such as never experienced before".

Water tables are also dropping world wide due to widespread and excessive pumping of fresh groundwater for agriculture, in large countries such as China, India and the United States, especially in arid regions. Fresh water resources are also declining in smaller countries like Algeria, Egypt, Iran, Mexico and Pakistan. A United Nations report indicates that the Himalayan glaciers, which are the primary water source for Asia's largest rivers, may disappear by 2035 due to rising temperatures. Approximately 2.4 billion people live in the watersheds of these great rivers that originate in the Himalaya, and countries such as India, China, Pakistan, Afghanistan, Bangladesh and Nepal may first experience unprecedented floods followed by severe drought for many decades. Ironically, we have an essentially unlimited supply of water on Planet Earth but 97 % of it is salt water and would have to be desalinated before use.

Famine is a chronic issue for much of Africa and Asia, and in 2005 the Famine Early Warning Systems Network designated Niger, Chad, Ethiopia, South Sudan, Somalia and Zimbabwe as presently being in "emergency status". And in 2006 the United Nations Food and Agriculture Organized issued a formal

76

warning that 11 million people in Somalia, Kenya, Djibouti and Ethiopia are in danger of famine due to a combination of drought and wars. At this time, the most serious crisis in Africa is the Sudan's Darfur region. Satellite imagery indicates that the Sahara Desert here is expanding at a rate of ~30 miles per year.

Droughts and famine are not restricted to Africa. They have occurred many times in Asia that is dependent on the Asian Monsoon to bring rainfall to the agricultural lands of India and China. Famines have also occurred in Europe. The Great Famine in Ireland (1845-1849) was in large part due to British rule and the fact that most farmland was owned by Anglican people of English descent - the "Anglo-Irish" The short term effect was the death of ~ 1 million people and another ~ 1 million refugees emigrating to the United States or Britain. Prior to "The Hunger", Ireland's population was more than 50 % of England's – today it is but an eighth. Presently the population of Ireland is ~ 6 million people, but there are more than 80 million people of Irish descent outside of Ireland - of which I am one.

Modern Food Crisis: Today we face yet another global famine. In 2007 and 2008 we experienced a dramatic rise in global food costs resulting in political, economic and social unrest in both developed and developing nations. The cause, or causes, of this most recent spike in food prices is highly debatable, in large part due to the complexity of our global agricultural system. Some have argued that climate related droughts, the high cost of petroleum, the use of bio-fuels, increasing demand by a growing human population, and the expanding global middle class have all contributed to the higher cost of food. In 2009, food prices began to drop, in large part due to the fall in the price of a barrel of oil. Many organizations, however, warn that this drop in food prices will only be temporary and that the world should not be lulled into "a false sense of security". The cost of food is likely to rise considerably in the near future.

Between 2006 and 2008 the mean global price for rice increased by 217 %, wheat by 136 %, maize (corn) by 125 %, and soybeans by 107 %. Many agricultural analysts point to a "perfect storm" of unprecedented issues to explain the increase in food cost. Factors such as bio-fuel production, growing demand for food, agricultural subsidies, climate change/crop disease, lower food reserves, changing global economy, reduced harvests, and oil prices have all have played a role.

The production of bio-fuels not only reduces food production, but also decreases arable land for food. One tank of bio-fuel gas produced from maize is about equivalent to an individual's annual consumption of corn in Africa – a factor referred to as "agflation". The World Bank concluded in 2008 that "...large increases in bio-fuels production in the United States and Europe are the main reason behind the steep rise in global food prices..." German Chancellor Angela Merkel, however disagrees with the conclusions of the World Bank. She cites "...poor agricultural policies and changing eating habits in developing nations..." as the primary cause. In 2008 former United States President George W. Bush stated that "...85 percent of the world's food prices are caused by weather, increased demand and energy prices..." and only "...15 percent has been caused by ethanol..."

Unprecedented human population growth has also been a contributing, if not *the* major, factor. Although population increase rates have dropped from 87 million people per year in the late 1980s to 77 million per year in 2007, this growth trajectory will result in a global human population of ~ 9 billion people by 2042, compared to our present population of ~ 6.5 billion people. That is a 27.8 % increase in human population and an additional 2.5 billion mouths to feed. Some, such as the writer/author Frances Moore Lappe have suggested that we do not have a food crisis, but warns that "...as long as food is merely a commodity in societies that don't protect people's right to participate in the market, and as long as farming is vulnerable to consolidated power off the farm, many will go hungry, farmers among them – no matter how big the harvest".

Idle farmland and government subsides also contribute to the food price problem. The *New York Times* reported on April 9, 2008 that in the United States alone, 8% of cropland lies idle, an area about the size of New York State. Agricultural subsidies that are used to pay farmers not to produce crops are currently ~ \$280 billion compared to only \$80 billion in 2004. Japan is also required to import some 767,000 tons of rice per year from the United States, despite the fact that Japan produces 121% of its consumption needs – the remainder is left to rot to use as animal food.

Climate and crop disease are other factors. A recent extended drought in Australia, the second largest wheat exporter in the world, resulted in a 39 % drop in national wheat production. Australia typically produces ~ 25 million tons of wheat per year, but due to drought, it produced only 9.8 million tons in 2006. Stem rust disease, caused by a virus, can result in 100% crop failure. Up to 85% losses of wheat harvest were reported in Kenya and Uganda that also spread across the Red Sea to Yemen, and is expected to spread to China and the Far-East. In 2009 parts of China' wheat producing region has experienced its worst drought in 50 years, and Chinese authorities have issued a "top level emergency" and have allocated $13 billion to the impacted area. Dr. Xu Yinlong, a researcher at the Chinese Academy of Agricultural Sciences has concluded that "This drought is occurring in front of the big backdrop of global warming and is part...of extreme weather events...it is definitely connected with climate change".

Environmental degradation of cropland is yet another factor in the decline of food production and its subsequent increase in price. Approximately 60,000 square kilometers (37,282 square miles) of agricultural land is lost per year due to soil erosion, pollution, water decrease and urbanization. Agricultural plants are also susceptible to pollution such as ozone generated by humans in the lower level of the atmosphere. The impact of ozone on the yield of food crops such as wheat and soybeans has been studied in the Yangtze Delta, China, where a 10-20 % reduction in biomass (plant weight) was found due to increased exposure to ozone.

To be certain they can feed their own people, countries such as China, Brazil, India, Vietnam, Indonesia, Cambodia and Egypt now have very tight bans on the export of rice. And, in North Korea, which is almost entirely dependent on foreign sources of food, an official has recently been quoted as saying that "...life is more than difficult...it seems that everyone is going to die...". And, according to the California Association of Food Banks, nearly all the food banks in the state are facing the "...beginning of a crisis..."

The consequence of food shortages is political, economic and social unrest. Food riots and/or government restrictions have recently been reported from Bangladesh, Brazil, the Cameroon, Egypt, Ethiopia, Haiti, India, Indonesia, Mexico, Mozambique, Pakistan, Panama, the Philippines, Russia, Senegal, Somalia, Tajikistan, and Yemen. And, a 2007 United Nations information fact sheet has projected a 49% increase in the cost of African cereals, and a 53% European increase.

A political summary of the current food crisis appeared in an October 23, 2008 *Associated Press* release that reported that "...former President Clinton told a U.N. gathering Thursday (October 16th) that the global food crisis shows 'we all blew it, including me', by treating food crops 'like color TVs' instead of as a vital commodity for the world's poor..." The former United States President suggested that "...decades of policymaking by the World Bank, the International Monetary Fund, and others, encouraged by the U.S...." resulted in "skyrocketing prices" and the decline of African food sustainability that forced many poor countries into even greater poverty.

This is the first food crisis since WW II, and the World Food Program estimates that 20 million of the poorest children are at risk. This "silent tsunami of hunger" has prompted world leaders, such as British Prime Minister Gordon Brown in April 2008 to call it a "...moral challenge...that is also a threat to the political and economic stability of nations..." Former UN Secretary General Kofi Annan stated that "...we are going through a very serious crisis and we are going to see lots of food strikes and demonstrations..." Also, Arif Husain, a food security agent for the World Food Program, summarized the problem very simply by stating that "...the human instinct is to survive, and people are going to so no matter what... if you get hungry, you get angry quicker...", and like Goodall's chimpanzees, humans will likely become more aggressive with an increase in the potential for global war.

Runaway Greenhouse: **Have you ever experienced an unusual cold spell during July in the Northern Hemisphere and overheard someone say – "well, I guess this disproves global warming"? Or, an unusual warm spell in January and someone says – "well, I guess this proves global warming"? The bottom line is that they are *both* wrong because they are reacting emotionally to changes in the weather, not climate change. Weather is the short-term, day-to-day variability we notice about the atmosphere. In contrast, climate is defined as the mean weather conditions of a region over a given, long-term period of time.**

Or have you ever engaged an educated, intelligent, articulate person who is 100% convinced that global warming is, or is not, occurring and when you ask them what they base their opinion on they have little, if anything, to say other than Rush Limbaugh or Al Gore said so? Well, I have, and I have learned that their "100 % opinions" tend to be based on their emotions, political persuasions, a lack of understanding of the Earth System, and limited scientific literacy (data).

Fortunately for us we have the IPCC, sanctioned by the United Nations, consisting of 1000s of climate experts from around the world. The IPCC has been evaluating the issue of global climate change since the

early 1990s, and in its most recent 2007 report concludes (among other things) that the scientific data from all continents, and most of the ocean, shows "...that many natural systems are being affected by regional climate change, particularly temperature increases..." and that "...it is likely that anthropogenic (human-induced) warming has had a discernable influence..." The only limitations of the IPCC are their reliance on computer models, rather than observations, and a need for political consensus.

A March 2, 2009 *Associated Press* article reports that global warming may well be worse than previously thought. Dr. Stephen Snieder at Stanford University and his colleagues reported that "...increases in drought, heat waves and floods are projected...to have adverse impacts, including increased water stress, wildfire frequency and flood risks starting at less than 1.8° F of additional warming above 1990 levels" projected by the IPCC. The report that appeared on-line from the *Proceedings of the National Academy Science* cited the 2003 heat wave in Europe that killed tens of thousands of people, and Hurricane Katrina, as examples of our "...increased vulnerability..."

From my personal experience the reason for all the misunderstanding, confusion and debate about global climate change stems from not knowing what the scientific definitions are and a general lack of scientific literacy about the Earth System, even among very well educated and articulate people. Probably the biggest misunderstanding is that the terms "greenhouse effect" and "global warming" are synonymous – they are not. By definition the "Greenhouse Effect" is the natural process by which all planets use to warm themselves. One has to realize that, with the exception of ultraviolet radiation absorbed by ozone in the stratosphere, the Sun's rays pass through our atmosphere with very little, if any, transfer of heat. The energy from the Sun heats the Earth's surface, and energy from the Earth is then reradiated back out to space as infrared radiation where some of it is absorbed by greenhouse gases – the greenhouse effect. For example, Venus, our closest neighbor to the Sun, has a very strong greenhouse effect and a surface temperature of ~ $460^{o}C$ ($860^{o}F$), whereas Mars, our closet neighbor farther from the Sun, has a very weak greenhouse effect and a chilly surface temperature of ~ $-55^{o}C$ ($-67^{o}F$). Some of this temperature difference is due to relative distance from the Sun, but it is largely a greenhouse gas concentration phenomenon.

In contrast, Earth's mean annual surface temperature of $15^{o}C$ (59°) is "just right" for life. Earth has a greenhouse effect of $33^{o}C$ ($91.4^{o}F$) – without it,

Earth's mean surface temperature would be a frosty -18°C (-0.4°F). Thus, we can conclude that the greenhouse effect is a "good" thing, because without the greenhouse effect the Earth, like Mars or Venus, would not be habitable for life as we know it.

Global warming, by definition, is simply an anthropogenic (human-induced) increase in the greenhouse effect. There are numerous greenhouse gases in the atmosphere such as carbon dioxide, water vapor, methane and ozone, among others. Carbon dioxide is neither the most abundant, nor the strongest greenhouse gas. It is, however, the one greenhouse gas mostly readily affected by humans – thus the focus on carbon dioxide.

When we use slash and burn techniques for purposes of deforestation, the burning of the trees releases carbon dioxide into the atmosphere. Similarly, when we pump oil and gas out of subsurface reservoirs that took millions of years to form, and we burn it at a much greater rate (~ 1,000,000 times), we also release carbon dioxide into the atmosphere. The Earth System, via a variety of processes (negative feedbacks) is trying to compensate for our rapid loading of carbon dioxide into the atmosphere, but on an annual basis the Earth System is only capable of removing about 50% of this anthropogenic carbon dioxide – the other half remains in the lower atmosphere for decades to centuries, where it absorbs outgoing energy from the Earth's surface and warms the planet.

If we want to know if global warming is really occurring all we have to do is look at carbon dioxide concentrations in the atmosphere over time. We can do this directly with instruments, or indirectly by using "proxy" data. There is an atmospheric chemistry station operated by the Scripps Institution of Oceanography at the top of Manu Loa, Hawaii that has been measuring the concentration of carbon dioxide in the atmosphere every 10 seconds since 1958. These unequivocal data – the Keeling Curve – document that since direct measurements began, there has been an increase in the concentration of atmospheric carbon dioxide from 315 parts per million to 387 parts per million, or an increase of 72 parts per million (19%) at a rate of 1.4 parts per million per year. We can conclude from these data that the Earth's lower atmosphere is out of "equilibrium" (steady-state or stasis) that would not be possible over such a short time period with a self-regulating Earth System or via geologic processes.

We also know via the measurement of air bubbles trapped in ice cores that now date back to ~ 1 million years, that our current value of carbon dioxide in the atmosphere (387 parts per million) is unprecedented for this time interval. *Over the past 1 million years we have gone through ~ 10 ice ages and carbon dioxide concentrations in the atmosphere never reached higher than 300 parts per million.* So, yes "Virginia", by definition, global warming *is* occurring! The most compelling evidence being the relentless rise of sea level due to the melting of land-based ice sheets and thermal expansion of the oceans as latent heat energy is added to the Earth System.

And many climate scientists agree that because of the "residence time" (amount of time prior to removal) of carbon dioxide in the atmosphere, we have already committed ourselves to climate change to at least the year 3000 even if we were to stop burning fossil fuels and deforestation practices immediately, which of course we are not doing. This statement is based on a recent publication in the *Proceedings of the National Academy of Sciences* by Dr. Susan Solomon (and her French and Swiss colleagues) who is at the National Oceanic and Atmospheric Administration's Earth System Research Laboratory in Boulder, Colorado. According to an *Associated Press* report on January 27, 2009 and a *Washington Post* article on February 2, 2009, Solomon's team concluded that "...climate change is slow, but it is unstoppable..." Solomon further states that "I think you have to think about this stuff more like nuclear waste than acid rain. The more we add the worse off we'll be. The more time we take to make decisions about carbon dioxide, the more irreversible climate change we'll be locked into".

The reason is that carbon dioxide "lingers" in the atmosphere much longer than other greenhouse gases – "...we can stop the clock, but we can't turn it back..." Our present level of carbon dioxide in the atmosphere is 387 parts per million and the United Nations, based on the latest IPCC consensus projections, has set a stabilization goal at 450 parts per million, but more realistic projections suggest a value of 550 parts per million by 2035, with a subsequent annual rise of 4.5 % per year. When carbon dioxide concentrations in the atmosphere reach ~ 600 parts per million, the southwestern United States, the Mediterranean, and southern Africa (home to billions of people) will experience severe droughts equal to, or greater than, the "Dust Bowl" years of the 1930s, and sea-level will increase a minimum of three feet.

An *Associated Press* story on February 15, 20009 reported "...humans are adding carbon into the atmosphere even faster than in the 1990s..." According to Dr. Christopher Field at the Carnegie Institution for Science "...carbon emissions have been growing at 3.5 percent per year since 2000, up sharply from the 0.9 percent per year in the 1990s..." He further went on to tell the *American Association for the Advancement of Science* that "...it is now outside the entire envelope of possibilities...considered in the 2007 report of the International Panel on Climate Change..." The largest factor appears to be the increased use of our cheapest fossil fuel – coal. "Past projections for declines in the emissions of greenhouse gases were too optimistic...no part of the world had a decline in emissions from 2000 to 2008..." In support of these statements, Dr. Anny Cazenave at France's National Center for Space Studies has satellite data that documents "...that sea levels are rising faster than had been expected..." – the key indicator that Earth is warming.

Dr. Mary-Elena Carr, associate director of the Columbia Climate Center in New York called these new findings "very sobering" and noted that even increased agricultural technology will not be able to cope with a "...finite amount of water and a growing population..." Solomon and her co-authors wrote that increasing sea-levels will result in "...irreversible commitments to future changes in the geography of the Earth, since many coastal and island features would ultimately become submerged..." In addition, the researchers noted that the world ocean has been storing about half of our carbon dioxide emissions for many decades, and that stored carbon dioxide will be released by warming (warm water can hold less carbon dioxide gas than cold water) in coming centuries even if we stop using fossil fuels today – a positive feedback.

U.N. policymakers tend to look only about 100 years into the future at a time in regard to global warming. New data call that policy into serious question because it is too short sighted. Dr. Ken Calderia at the Carnegie Institution's Department of Global Ecology has argued for 500 year projections into the future, noting that politicians tend to look at the more uncertain aspects of global warming. And Dr. Carr warned "...that the parts that we don't know, that are possible but very uncertain, shouldn't get in the way of what we do know..." Dr. Alan Robock of the Center for Environmental Prediction at Rutgers University agrees saying that carbon dioxide in the atmosphere is "...not like air pollution where if we turn off a smokestack, in a few days the air is clear..."

Key questions are whether or not technology will, can, or should "save" us from global warming. And, is there a real possibility that we may be causing a "runaway greenhouse effect"? In 2009 climate experts Dr. Michael Mann and Dr. Lee Kump, both at the Pennsylvania State University, published a wonderfully illustrated guide to the "dire predictions" of global warming from temperature increases, melting of ice caps, rising sea levels, droughts, floods, ecosystem collapse, war, famine, pestilence and death. And they quote Dr. James Lovelock, from his 2006 book *The Revenge of Gaia*, as saying "...for now, the evidence coming in from...around the world brings news of an imminent shift in our climate towards one that could easily be described as Hell: so hot, so deadly that only a handful of teeming billions now alive will survive". Pretty scary stuff!

Rather than discuss all the potential specifics about global warming I will focus on the possibility of a "runaway greenhouse" that might produce "Hell on Earth". Nobody can *predict* the future, and if they tell you they can, do not believe them. We can, however, make *projections* based on data in hand and our current knowledge and understanding of the Earth System. The IPCC has made projections for the future based on a number of different scenarios. Such projections have been conducted since the pioneering work of Dr. James Hansen, a climatologist and Director of NASA's Goddard Institute for Space Studies, and a member of the National Academy of Science. There is considerable uncertainty in any projection because we do not know how global society will respond in the future, and we do not yet fully understand the entire Earth System.

Deforestation, an anthropogenic process that releases carbon dioxide into the atmosphere, has been occurring for centuries. Native Americans and our own ancestors deforested much of North America, as was much of Europe and Asia that led to today's economically developed countries. It is important to realize this because today's focus on deforestation is the tropical rainforests located mostly in developing countries.

National Geographic refers to deforestation as a "modern-day plague". It is the clearing of Earth's forests on a massive scale that damages the quality of the land. Forests cover about 30 % of the Earth's surface today, but "swaths the size of Panama are lost each and every year".

So, why do humans deforest the landscape? Because of greed and the want of money or the necessity to support one's family. The largest culprit is for agricultural purposes – more room for raising crops and grazing cattle – usually by "slash and burn" techniques. Logging for the world's supply of wood and paper, as well as urbanization are also reasons why we cut down our forests. In addition to releasing carbon dioxide, deforestation also reduces the number of trees that can absorb carbon dioxide from the atmosphere - a positive feedback - and destroys habitats for millions of species. Forests are, however, a potentially manageable, renewable resource if our global society decides to do so, unlike the use of fossil fuels that is not.

Despite our extensive use of fossil fuels to power the global economy over the past three centuries, there are enormous amounts of fossil fuels still available, yet finite. Fossil fuel *reserves* have already been discovered and are immediately available for use, whereas fossil fuel *resources* are those that geologists believe exist but have yet to be discovered or tapped into. Our total fossil fuel reserves will increase in the coming decades as technological advances allow previously unknown or unrecoverable resources to be found. Presently, fossil fuels include oil, natural gas, coal, oil sand, oil shale, and methane clathrates.

At the beginning of the 21st century world reserves of oil were estimated to be 1.3 trillion barrels, and by 2003 oil consumption was 29 billion barrels of oil. Natural gas reserves were estimated to be about 6,000 trillion cubic feet with a global consumption rate in 2003 of 96 trillion cubic feet per year. And, coal reserves were estimated at 1.1 trillion tons with a 2003 consumption rate of 5.5 billion tons. Assuming the same consumption rate as for 2003, simple math tells us that we have a 45 year reserve of oil, a 63 year supply of natural gas, and a 2000 year supply of coal. And, remember, these are *reserve* numbers.

Resource values may be considerably higher considering the massive volumes of methane known to be in the deep sea and in our vanishing global permafrost. The current resource estimate for ocean clathrates (ice that contains a large amount of methane within its crystal structure) alone is between 35 and 177 quadrillion (1,000,000,000,000,000) cubic feet! Thus, we have an ample reserve of fossil fuels and an even larger resource of fossil fuels to burn in the future that will, if burned, release massive amounts of carbon dioxide into the atmosphere.

Future emissions of carbon dioxide in billions of tons per year are highly uncertain. Thus, the IPCC has four basic "storylines" representing a group of emission scenarios: 1) A1 – one global family; 2) A2 – a divided world; 3) B1 – global utopia; and, 4) B2 – local utopia.

The largest projection is the A1 fossil-fuel-intensive scenario (formally known as "business as usual") that estimates fossil fuel emissions at the end of the 21st century to be about 35 billion tons per year, compared to 21.3 billion tons per year today. This scenario assumes substantial reduction of regional differences in per-capita income, rapid economic growth, peak population in 2050 declining thereafter, and the rapid introduction of new, more efficient technology. The lowest projection is the B1 storyline of "global utopia" that projects fossil fuels emissions by the end of the 21st century to be between about 5 and 10 billion tons of carbon. This scenario assumes an emphasis on global solutions to sustainability, rapid change to information and service economies, peak population in 2050 declining thereafter, reduction in intensity of demand for materials, and introduction of new, clean, efficient energy technology. No one really knows which one, if any, of the IPCC storyline projections will turn out to be correct. We could be emitting 35% more carbon by 2100 or about 22 % less. It is important to realize, however, that unless we *stop* using fossil fuels completely we will continue to load the atmosphere with more energy absorbing greenhouse gas, regardless of the scenario, well past 2100.

One scenario generated by computer modeling a few years ago and presented in Dr. Lee Kump and his colleagues' book *The Earth System (2nd edition)* assumed a fossil fuel reserve of 4.2 billion tons of carbon that is totally consumed within the next 400 years, as well as assuming that deforestation trends continue only until there is 30 % forest coverage. The results project a very rapid rise in the concentration of atmospheric carbon dioxide from pre-industrial levels of 280 parts per million to as much as 2,100 parts per million by the year 2300. If this actually occurs the mean annual global surface temperature of Earth would nearly double, increasing to 28.5°C (83.3°F) from our present value of 15°C (59°F).

Clearly, such a scenario would result in a "runaway greenhouse effect" that would wreak havoc on human population and global ecosystems. The computer model, however, does not include the role of potential *positive* feedbacks that *amplify* change within the Earth System, of which there are many. One of the best examples of a positive feedback is that as the Earth warms, the white colored ice sheets in Greenland and Antarctica melt and are replaced by green plants, water or dark soil/rock. Light colors (ice) reflect more solar energy back out to space before it can heat the Earth's surface than darker colors. Thus more solar energy would be absorbed by the Earth's surface, further increasing mean surface temperatures, particularly at high latitudes.

Another example of a positive feedback comes from a 2009 research paper published in *Science*. Researchers found that trees in old-growth forests in the mountains of the western United States are dying faster than ever before. In fact over the past 17 to 37 years the death rate of mature trees has doubled.

Their deaths have been linked to rising regional temperatures as a result of global warming. As temperatures increase there is earlier snowmelt, drought, forest fires, and infestations by insects. Dr. Mark Harmon at Oregon State University stated that "...the ultimate implications for our forests and the environment are huge..." because as the forests shrink they loose their capacity of removing carbon dioxide from the atmosphere which accelerates global warming – a positive feedback. Dr. Nathan Stevenson of the U.S. Geological Survey's forest ecology center said that "...the droughts are lasting longer, and they're helping all those things that like to eat trees". The scientists also found that where tree death rates have accelerated, fewer new trees are replacing the dead ones.

Thus, a modern runaway greenhouse scenario is certainly possible, and with positive feedbacks could be significantly worse than the scenario presented by Kump and his colleagues!

Permian Mass Extinction: Have there been times in the geologic past when a "runaway greenhouse effect" occurred? The answer is yes, and at multiple times. Rather than describe the details of each of these runaway greenhouses in Earth history, I have chosen the most dramatic event that has the greatest applicability to this book – The End Permian Mass Extinction – informally known as the "Great Dying". The Permian extinction occurred 251.4 million years ago, within an interval of only ~ 10-60 thousand years, and is considered to be the most severe mass extinction in Earth history. It resulted in the extinction of up to 95% of all marine species, 70% of terrestrial vertebrate species, and the only known mass extinction of insects – the "mother of all mass extinctions".

There have been a number of proposed mechanisms for this cataclysmic event over the years ranging from gradual environmental change to an instantaneous asteroid impact – similar to the impact that killed off the dinosaurs 65 million years ago. But the current consensus amongst geologists is that the Permian mass extinction was a consequence of a "runaway greenhouse" event, as the observed pattern of the extinction is consistent with "hypercapnia" - excessive levels of carbon dioxide. Too much of most any chemical can be lethal. We need water to survive, but if we inhale too much we drown. Excessive amounts of carbon dioxide are toxic because they reduce the ability of respiratory systems to supply oxygen, make body fluids more acidic, and can cause narcosis ("intoxication").

The timing of the Permian mass extinction is consistent with the age (251.2 +/- 0.3 million years) of the so-called "Siberian Traps" that are a product of the largest known volcanic event in Earth history covering more than 77.2 million square miles with lava. These largely effusive lavas also released large volumes of carbon dioxide into the atmosphere that led to a temperature increase of up to 8.1°F. Such warming was enhanced by a positive feedback that released large quantities of methane from the ocean floor and permafrost that is 22 times stronger as a greenhouse gas than carbon dioxide. The result was a "runaway greenhouse" effect that took the Earth System at least 4-6 million years to recover from.

Another positive feedback is that deep-sea circulation ceased, causing the oceans to go "anoxic" (devoid of oxygen) and poisoned by hydrogen sulfide gas. Organisms died because of excessive amounts of carbon dioxide, overheating, anoxia and/or hydrogen sulfide poisoning. In a 2003 research article on "Mass Extinction" published in *The Traprock* (v. 2) Dr. Stephen Goldberg and his colleagues warned of a possible future runaway greenhouse mass extinction initiated by global warming stating that "...another mass extinction could be in sight in the foreseeable future. If that were the case men (humans) could very likely be the victims of one of the biggest crises on Earth during its existence".

Next Ice Age: Syndicated columnist George Will reminded us on February 15, 2009 that "Global cooling was (a) hot topic, too". During the 1970s "...a major cooling of the planet was widely considered inevitable..." and that "...a full blown 10,000 year ice age...involving extensive Northern Hemisphere glaciation..." would ultimately occur. At the time "...meterologists were almost unanimous the trend will reduce agricultural productivity...triggering catastrophic famine..." Will cites Montaigne's axiom that "...nothing is so firmly believed as what we least know..." to support his claims that "...real calamities take our minds off hypothetical ones..." such as global warming. In my opinion, George Will is both correct and wrong at the same time.

When you examine the existing geologic and paleoclimate data closely, as I have, you find the following. Over about the past 1 million years, large ice sheets in the Northern Hemisphere have advanced and retreated about 10 times, or once every 100,000 years. These so-call "Pleistocene" ice age cycles are a culmination of a ~ 50 million year long cool down of the Earth as a result of the plate tectonic collision of India with Asia to produce the Himalaya Mountains and the Tibet Plateau. This enormous collision either reduced Earth's plate tectonic "machine" resulting in less transfer of carbon dioxide from the lithosphere to the atmosphere, and/or it exposed so much fresh rock that chemical weathering, which requires the use of carbon dioxide, increased significantly.

Although the ice ages are a culmination of a 50 million year cool down of Planet Earth, their cyclic nature is tied to subtle changes of Earth's orbital parameters such as it axial tilt, shape of its orbit around the sun, and the timing of seasons on that elliptical orbit. These orbital parameters change slightly and have but a small forcing effect on Earth's climate. However, we need to always think of Earth as a holistic, integrated *system* that has numerous feedbacks. Ice sheets advance when Earth's orbital parameters result in cool summers at high latitudes in the Northern Hemisphere that are then enhanced by numerous positive feedbacks that amplify weak forcing by the orbital parameters. Ice sheets retreat when the orbital parameters result in warm summers (snow and ice melt) at high latitudes in the Northern Hemisphere amplified by positive feedbacks. The reason for the 100,000 year cycle of glacial advance and retreat over the past ~ 1 million years is a result of Earth's orbital parameters being controlled by solar system-scale gravity over long intervals of time.

We have excellent data on this 100,000-year period of glacial advance and retreat from both deep ocean sediments and ice cores. Glacial periods build slowly, but interglacial periods (like the one we are presently at the end of) occur abruptly and last for as little as ~10,000 years. The most recent glacial period began ~125,000 years ago and reached its maximum extent ~21,000 years ago. At the time of the last glacial maximum major cities like Moscow, London, Dublin, Boston, New York City, Chicago, Seattle, all of Canada and much of northern Eurasia were covered with glacial ice up to ~ 1+ mile thick. The area that the last great ice sheets covered is now home to billions of people who would have to be displaced to the south if the ice advances yet again, resulting in tremendous overcrowding and an enormous reduction of available food resources. Billions of people might die due to war, disease, and/or famine. There would be total chaos in global politics and the global economy.

Unfortunately for us, we are at the end of our present warm interglacial period. In fact, about 5,000 years ago Earth's orbital parameters started to change and began a plunge into the next ice age that should peak in a little less than 100,000 years. Many climate scientists believe that the Little Ice Age (16th century to 1850) was the first significant step of the

Earth System into another ice age. Yet, our instruments and observations are telling us that warming, not cooling, is occurring throughout much of the world at an increasing rate.

Global warming is, by definition anthropogenic, but we are part of the natural Earth System. This brings up a very interesting philosophical issue. Is global warming "bad" or "good"? We have all heard and read about the "dire predictions" of global warming. But there are those that potentially see global warming as a natural response of the Earth System to stem the tide of the next ice age - again, no one can predict the future.

VI) STORYLINES

I hope by now that you realize that the Earth System has numerous "weapons" in its arsenal to deal with the potential of human population, aided by technology, exceeding the carrying capacity for humans on Planet Earth.

So, how will the Earth System respond? I do not profess to have the answer to that question, but there are a number of possibilities that I would like to explore with you, both in verse and with a few simple graphs. I will adopt the approach of the IPCC and use "storylines" (scenarios) of which there are many, but I will focus on the major end-member examples.

Norm: Populations of any type of living organism are a group of individuals of the same species living in the same geographic space. Most populations ("the

NORMAL STORYLINE

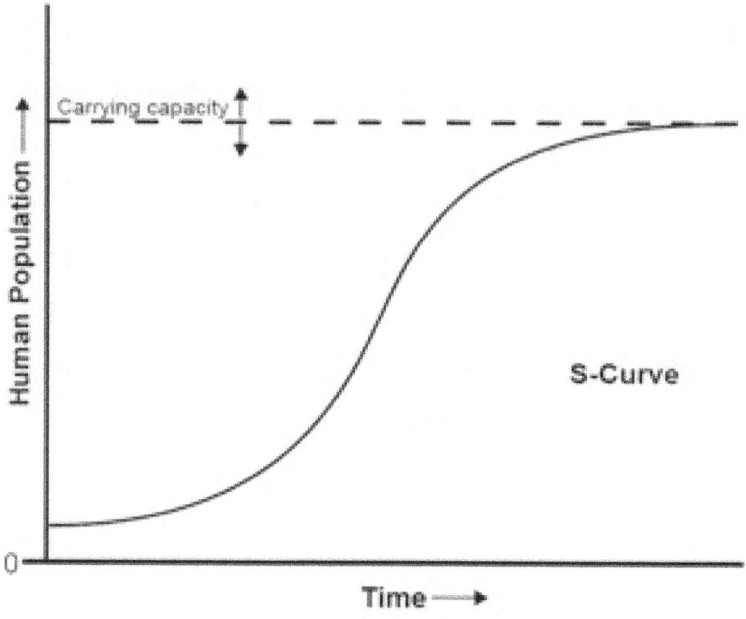

norm") have three cycles: growth, stability and eventual decline. Population growth takes place when necessary resources are greater than the rate at whicha population can use them. Reproduction tends to be rapid and death rates are low resulting in population growth. Stability occurs when the growing population eventually reaches the level of its necessary resources. Decline is a reduction in the numbers of a population, and can eventually lead to extinction over long periods of time.

Natural checks (negative feedbacks) on population growth are divided into "density-independent" factors and "density-dependent" factors controlled by population per unit area of a species. Density-independent checks are simply the "vagaries of the physical environment" for a species, such as drought, freezes, storms, floods and wildfires. Such checks not only limit population growth but may also reduce it, and exert their influence independent of population density. From an Earth System Science perspective Planet Earth is an "island" in our solar system, galaxy and universe – it is our habitat. Catastrophic declines in populations are particularly susceptible to "island-based" populations, such as Earth from a large-scale perspective.

Density-dependent checks on population growth result from competition between members of the same species, as they begin to compete for finite resources, such as food. The larger the population density of a species the stronger the density-dependent check on population growth will be, unless the process is somehow intervened – for example by technology. Other types of density-dependent checks include the requirement of an ecological niche (for humans it is our entire planet), reproductive competition, migration, parasitism, and predation. Predation, however, does not readily apply to humans as we are now at the top of the food chain.

Properties of a population include size, density, dispersion, demographics, growth and growth limits. "The characteristics of a population are shaped by the interactions between individuals and their environment on both ecological and evolutionary time scales, and natural selection can modify these characteristics of a population". The simplest type of population growth is when there are no limitations to increasing numbers of a species. Intrinsic growth rate becomes exponential. However, in all environments/habitats, including Planet Earth, a population is never able to achieve true unlimited growth due to finite resources and negative feedbacks that limit the maximum population number.

Many species populations exhibit "logistic (restricted) growth" that is a "sigmoid" or a S-shaped curve - the "norm". When population is low and resources high, growth initially starts slowly but soon becomes exponential. As resources begin to decline with exponential population growth, growth rate decreases as it approaches some critical limiting requirement – its "carrying capacity". Stability becomes established as the population growth rate stabilizes in response to its environmental carrying capacity. With multiple recycling systems in the Earth

System, population stability ("the norm") may persist for long periods of time. There are species living on Planet Earth today, that existed many millions or even billions of years ago. "Over the long term, many populations remain fairly stable in size and are presumably close to carrying capacity that is determined by density-dependent factors. Superimposed on this general stability, however, are short-term fluctuations due to density-independent factors".

Equilibrium: Unfortunately, the human population growth curve has not followed the normal storyline S-

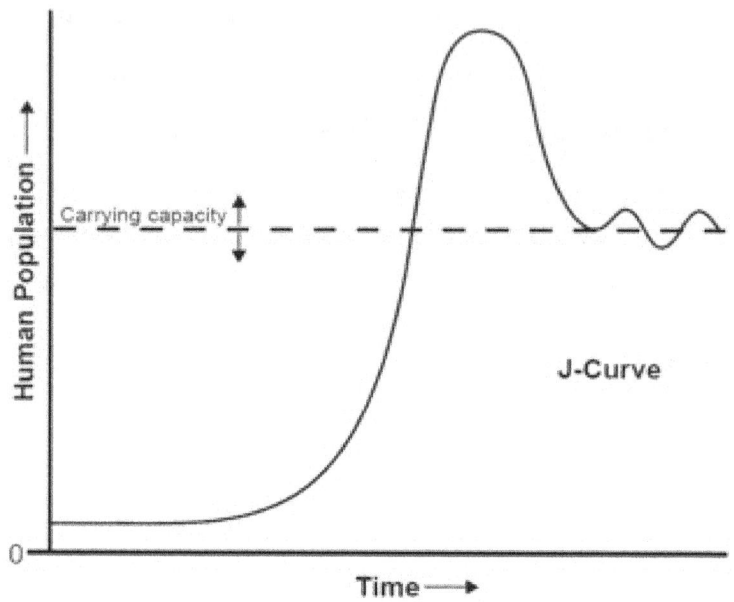

shaped curve. Ours is an exponential J-shaped curve that is presently out of control and in crisis. The primary reasons for this are our intellect as well as technological advancements in medicine and agriculture. Once a component of the Earth System, such as human population, gets out of "equilibrium" (stability) the Earth System will eventually impose negative feedbacks in an attempt to regain equilibrium.

The Earth System has numerous "weapons" at its disposal to impose negative feedbacks and obtain equilibrium. In the "equilibrium storyline" human population greatly exceeds its carrying capacity on Earth and a variety of negative feedbacks reduce the human population to its natural carrying capacity level. One may look at this storyline as "good", because no population can grow exponentially forever, and it would leave an equilibrium number of people on the planet to perpetuate our species.

The idea that excess human population can somehow be transported to other bodies in our solar system, such as the Moon or Mars is, in my opinion, highly untenable. The literature on this subject began in 1869 with Edward E. Hale's writings about "...inhabited artificial satellites..." and continues to modern times with Harrison Schmitt's 2007 book *Return to the Moon.*

There are good reasons why no humans live on other solar system bodies – they are not suitable habitats for human life! The concept of space colonization – the autonomous, self-sufficient human habitation of locations outside Earth – however, has been a major theme of *science fiction* and, unfortunately, numerous national and international space programs. In 2005, then NASA Administrator Michael Griffin stated that: "...the goal isn't just scientific exploration...it's also about extending the range of human habitat out from Earth into the solar system as we go forward in time...in the long run a single-planet species will not survive...if we humans want to survive for hundreds of thousands or millions of years, we must ultimately populate other planets...I know that humans will colonize the solar system and one day go beyond". At the present time we do have an International Space Station that provides a "permanent" human presence in space, but it is not an autonomous, self-sufficient structure.

To establish human colonies beyond Planet Earth there are numerous requirements that, in my opinion, will never be fully met – even with advanced technology. Colonies would require food, space, construction materials, energy, communications, life support, simulated gravity, and solar radiation protection, such as Earth's ozone layer. And humans require oxygenated air, water, food and reasonable temperatures that would have to be supplied or recycled to avoid a catastrophic "crash" of a space colony. The Biosphere 2 experiment attempted to simulate a space colony on Earth but experienced multiple problems, the most serious of which was that oxygen had to be replenished externally.

Many solar system localities have been suggested as potential sites for space colonies including Mars, our Moon, other planet's moons, Mercury, Venus, gas giants, as well as free space. All of these suggestions have very severe limitations in terms of supporting space colonies, primarily the basic life support needs of humans. Although the famous physicist Dr. Stephen Hawking has said that: "...the long term survival of the human race is at risk as long as it is confined to a single planet. Sooner or later, disasters such as an asteroid collision or nuclear war could wipe us all out. But once we spread out into space and establish independent colonies, our future should be safe. There isn't anywhere like the Earth in the solar system, so we would have to go to another star". With all due respect to Professor Hawking, I have to disagree. I believe we should be focusing our intellect, financial resources and technology on developing a sustainable Earth, and reducing our numbers, rather than spending billions and billions of dollars on space exploration for the purpose of potential human colonization.

Catastrophe: If James Lovelock is correct that the Earth System views us as a "pest" whose population numbers need to be dramatically reduced, it does have the "weaponry" to do so. Whether it is nuclear winter, superbugs, famine, runaway greenhouse and/or the next ice age, there is the potential for a catastrophic "crash" in the number of humans living on Earth. In this storyline our J-shaped human

113

population growth greatly exceeds Earth's carrying capacity for humans and population numbers are then greatly reduced via negative feedbacks, to a level

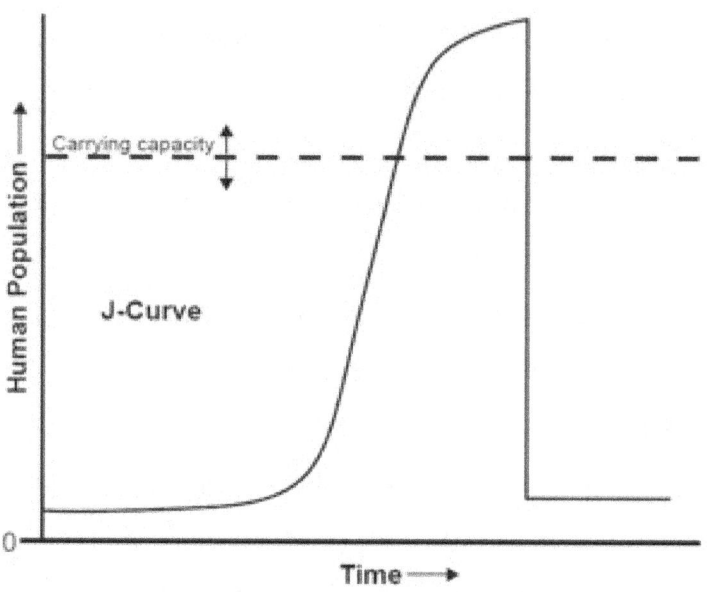

of sustainability prior to the scientific, medical and technological revolutions of the 19th and 20th centuries.

There are numerous examples of population crashes that have been verified by biologists over the years, in fact far too many to discuss all of them here. Rather, I will use a small selection of examples, to illustrate the point. In 1962 Rachel Carson wrote her famous book *Silent Spring* that is commonly credited to have inspired the modern environmental movement. In it she wrote – "It was spring without voices. On the mornings that had once throbbed with the dawn of the chorus of robins, catbirds, doves, jays, wrens, and scores of other bird voices, there was now no sound; only silence lay over the fields and woods and marsh". She was referring to the dramatic population decline of many songbirds – up to 90 % - during the past 50 years, a trend first noticed by John James Audubon over 100 years ago. Carson brought to light the effect of pesticide spraying (such as DDT) has on the population of songbirds. Although DDT is now banned in the United States, it is still sold and used in other countries, such as in Central and South America, which are winter migration grounds for many songbirds.

When organisms have their habitats disturbed or destroyed they get M.A.D. – move, adapt or die. Habitats naturally undergo continuous change over long periods of time as the ambient climate changes. There are "geographic" changes resulting from erosion and earthquakes, "short-term" changes caused by droughts or fires, and "human-influenced" changes related to habitat destruction or global warming. One of the most significant threats to songbird populations is the loss of habitat in their winter migration regions.

A second example is the "southern resident orca population crash" in 2008. A meeting organized by the Puget Sound Partnership at the Friday Harbor Laboratories that included the National Marine Fisheries Service, NOAA, and other scientists came to the conclusion that the crash in the orca (killer) whales was starvation. The orcas' diet consists of about 70% king (Chinook) salmon that had experienced a crash earlier in the year off California. While a lack of food was the primary cause of the orca population crash, it was amplified (positive feedback) by motorized sight-seeing boats. It is known, by measuring respiration rates, that power boats increase whale metabolism requiring more food. Whales also swim faster, dive deeper and travel farther when harassed by power boats, further increasing their need for more food. And, whale sonar used for hunting is disturbed up to 97% in the presence of motorized boats. The Endangered Species Act and the Marine Mammal Protection Act state that it is illegal to "pursue" endangered species which brings up the definition of "pursuit". In a 2008 article in the *Island Guardian* Mark Anderson defines pursuit as "…any commercial entity advertising certainty of finding and seeing whales for paying customers…."

And my final example comes from a scientific paper by Dr. Daniel Simberloff and Dr. Leah Gibbons in the Department of Ecology and Evolutionary

Biology at the University of Tennessee entitled – "Now you see them, now you don't! Population crashes of established introduced species" – published in 2004 by the journal *Biological Invasions*. This paper has applicability to our previous discussion about human space colonization because if we were to colonize another planet we would be an "invasive species". These researchers found that "...substantial populations of invasive non-indigenous species occasionally collapse dramatically..." Disease is often cited, but not proven, as the cause of such collapses that have rarely been studied, and some remain a mystery. The researchers studied an invasive snail and pondweed that underwent "...rapid expansion followed by rapid decline..." They suggested that their observations may simply be a "boom-or-bust cycle", but they pointed out that in restricted habitats, such as an island (ex. - Earth in the solar system), that recovery from a crash might never occur, and lead to extinction. "Spontaneous collapse" of organism populations has repeatedly been observed by biologists. But then again, some invasive species in the northeastern United States, such as the zebra mussel and Eurasian milfoil (lake-weed), are presently doing extremely well in their "new" homes.

Extinction: Extinction is defined as the "death of every member of a species or group of organisms". Species extinctions have occurred throughout much of Earth's history, the most famous one about 65 million years ago, when an asteroid hit the Yucatan Peninsula of Mexico and wiped out the dinosaurs.

Species become extinct when they can no longer adapt to changes in their environment. For a typical species this occurs within about 10 million years after its first appearance. But we are an unusual species. We have existed for less than 0.0005 % of Earth history, but have rapidly developed into the dominant species on our planet due to our intellect and technology. But remember that 99.99999....% of all species that have ever lived on Planet Earth before us,

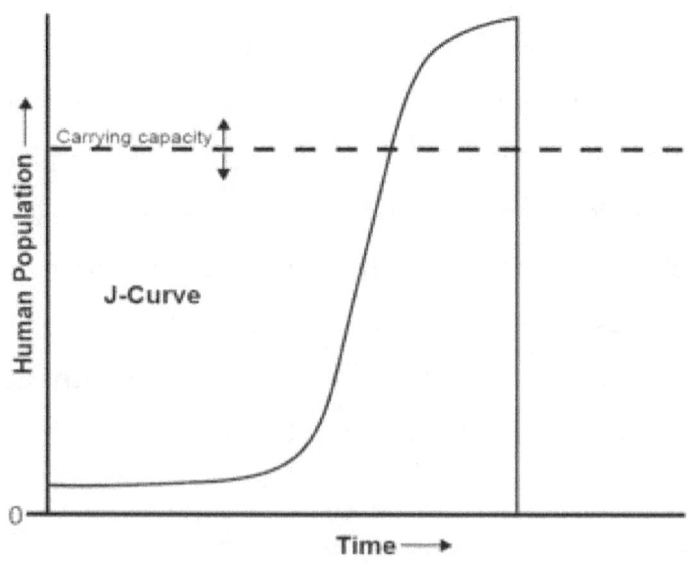

are now extinct.

Species have come and gone on Planet Earth for at least the past 3.8 billion years – it is a very natural process of Darwinian evolution. However, over the past 540 million years there have been five major mass extinctions including the above mentioned Cretaceous-Tertiary extinction 65 million years ago, and the end-Permian extinction discussed previously.

And, many paleontologists and biologist believe that we are presently in the midst of a sixth mass extinction – the Holocene extinction – that projects up to 50% of all living species will become extinct by the end of this century! This Holocene extinction event started about 11,000 years ago, and has been linked to processes in both the biosphere and the atmosphere. Dr. Peter Ward, Professor of Biology and of Earth and Space Studies at the University of Washington in Seattle, and Dr. Daniel Simberloff, Professor of Environmental Science, Ecology, and Evolutionary Biology at the University of Tennessee in Knoxville, both agree that this sixth mass extinction in Earth history is a combination of overhunting by humans, habitat destruction and climate change.

Conservationists have recently (2008) catalogued the world's mammal population (including humans) for the first time in more than a decade. Dr. Jan Schipper of the International Union for the

Conservation of Nature (IUCN) in Switzerland reported that "...our results paint a bleak picture of the global status of mammals worldwide...one in four species is threatened with extinction and...the population of one in two is declining..." The report was co-authored by more than 100 scientists, and Dr. Julia Marton-Lefevre, also of the IUCN, added that "...within our lifetime hundreds of species could be lost due to our own actions..."

The IUCN's "red list" of threatened mammals consists of approximately 45,000 species of which about 17,000 are in "near-term" danger. About 3,000 species are listed as "critically endangered", about 5,000 are "endangered" and approximately 9,000 are "vulnerable to extinction". Since 1500 about 76 mammalian species have gone extinct. Primates (our ancestors) are in particular danger because they are used for "bush meat" in Africa, and suffer major loss of habitat in Southeast Asia due to deforestation.

Larger mammals, such as humans, are more susceptible to extinction because they "...tend to have lower population densities, grow more slowly, and have larger home ranges..." Overkill hunting and deforestation of tropical rainforests are impacting land mammals, and accidental deaths, pollution and disease are reducing the number of marine mammals.

Climate change is also a factor – for example, the loss of sea-ice habitat for polar bears and harp seals.

Between 2006 and 2008, two new "unifying" hypotheses have been proposed for mass extinctions in general. The "press/pulse" hypothesis of Mr. Ian West and Dr. Nan Crystal of Hobart and William Smith Colleges in Geneva, New York rejects "...the all-or-nothing approach to mass extinction, calling instead on a combination of deadly sudden catastrophes – pulses – with longer, steadier pressure on species – presses". The researchers presented their hypothesis to the *Geological Society of America* based on a database of marine organisms that have gone extinct over the past 488 million years.

A second, and potentially more applicable idea was published in 2008 by Dr. S.A. Wooldridge at the Australian Institute of Marine Science in Townsville entitled – "Mass extinctions past and present: A unifying hypothesis" – in *Biogeosciences Discuss.* The author argues that enzymes in living organisms are "agents of life" because they only function within a narrow band of environmental conditions, such as temperature and pH. pH is a relative measure of acidity that ranges from 0-14 with 0 being extremely acidic, 7 being neutral, and 14 being very non-acidic. Wooldridge hypothesizes that the pH change of a single enzyme urease (an enzyme that catalyzes the addition of water in urea) may be the "unifying kill-mechanism" of extinctions.

Triggering the dysfunction of urease can be the result of gradual and/or catastrophic environmental change that results in organisms entering "dead zones". Wooldridge's research suggests that the threshold or "tipping point" for the dysfunction of urease is a pH = ~ 7.9. For most terrestrial and marine environments, the pH value of 7.9 corresponds to a carbon dioxide atmospheric concentration value of ~ 560 parts per million which is double the pre-industrial value of 280 parts per million, and only ~ 173 parts per million from today's value of ~ 387 parts per million. "The urease hypothesis...predicts an impending Anthropogenic extinction event...unless future CO_2 levels can be stabilized well below 560 parts per million". At our current rate of atmospheric carbon dioxide increase, 560 parts per million "...may be exceeded as early as 2050"!

In the extinction storyline human population greatly exceeds the carrying capacity for our species on Earth, and the Earth System responds using negative feedbacks such as nuclear war, pandemics, famine, runaway greenhouse effect, and/or the next ice age and violently results in the rapid extinction of *Homo sapiens* – never to be resurrected on Earth again!

122

Which one, if any, of these four storylines occurs - I honestly cannot say. But if I were a betting man, that I am, I would place my money on the extinction storyline. Why? Human nature and statistical probability!

Historical Analogs: Dr. Jared Diamond, a professor of geography and physiology at UCLA published a book in 2005 – *"Collapse: How Societies Choose to Fail or Succeed"* – in which he states "…readers should learn from the past…" He suggests that societal collapses have been a combination of the environment, climate change, hostile neighbors or trade partners, and/or societal responses. He, as do I, argues that the *root* of societal (system) collapse is human *overpopulation* relative to the Earth systems' *carrying capacity*. So, let us take a brief look at the collapse or extinction of a few selected past human civilizations - systems.

The Maya, or Mesoamerican civilization was the only known civilization in the western hemisphere to have a fully developed written language before Columbus "discovered" America in 1492. This civilization was initiated ~ 2000 BC with its highest level of prosperity between 250 and 900 AD. At its peak, it was one of the most densely populated and culturally sophisticated societies in the world. The

Maya system collapsed during the 9 century. Geologist and paleoclimatologist Dr. David A. Hodell, Woodwardian Professor of Geology at the University of Cambridge (formerly at the University of Florida) and his colleagues, have studied lakes on Mexico's Yucatan Peninsula and discovered that there was a devastating drought during the 9th century that caused the collapse of the Maya civilization that was an agrarian (agriculture) based society – what a BBC television series referred to as an "Ancient Apocalypse" – and what I would refer to as an analog for the "catastrophe" storyline.

We can also use the fall of the Roman Empire as another analog of system failure. The Roman Empire was an autocratic system that controlled large tracks of land, at its peak ~ 2.3 million square miles. The Empire began with the appointment of Julius Caesar as perpetual dictator in 44 BC. The traditional date for the fall of the Roman Empire is 476 AD. Numerous hypotheses have been proposed as the cause of the fall of the Roman Empire, but most scholars today believe it was a complex "transformation" – that I would liken to the "equilibrium" storyline.

The demise of the British Empire is another example of an "equilibrium" storyline. This Empire consisted of dominions, colonies, protectorates,

124

mandates and other territories governed by the United Kingdom. It originated in the 16th and 17th centuries and at its peak was the largest empire in history. In 1922 the British Empire controlled ~ 458 million people, about 25 % of the world's human population at that time, as well as more than 13 million square miles of land, or approximately 25 % of the Earth's land surface.

The growth of Germany and the United States eroded the Empire's economy by the end of the 19th century, and hostility between Britain and Germany led to the start of WW I. After this war the British Empire was no longer the strongest industrial or military power in the world. Two years after WW II, Britain gave independence to its former colony, India, and during the remainder of the 20th century the Empire began a program of global decolonization. As of 2002, the British Empire has been know of as the British Overseas Territories, which consists of a "...handful of mostly small islands with low levels of population...", with the lone exception of Northern Ireland. Today, the British government is responsible for the military safety of fourteen independent countries, such as Bermuda, the Falkland Islands, and the Turks and Caicos Islands in the Bahamas archipelago.

And, finally, we can use the former Soviet Union as an analog for the "extinction" storyline because the Soviet Union no longer exists. The collapse of the Soviet Union occurred at the end of the 1980s. The primary cause was economic due to the Soviets losing the arms race combined with international economic competition. Soviet people started to turn to imports for their goods and services. Other factors were political and psychological such as the war in Afghanistan and the nuclear debacle at Chernobyl. Another factor was a "culture of war" that produced a lack of honest information and propaganda. A 1991 article stated that "Secrecy and restricted movement, the hallmarks of militarism and bureaucracy, pervaded Soviet society...hampered scientific institutes... (and) all levels of the system...were isolated from each other..." Gorbachev began to transform the Soviet's military industry into one that provided products for civilians. All of these factors led to a "...profound alienation of the Soviet people..." and its ultimate "extinction".

VII) EPILOGUE

The life expectancy of any planet in the Universe is a function of its sun (star). Each star is unique and its lifespan is dependent upon its mass - the larger the mass, the shorter its lifespan. Stars are "born" via gravitational attraction of interstellar dust and gas until they reach a "critical mass" at which time the process of nuclear *fusion* begins. Our nuclear power plants generate electricity by nuclear *fission*, or the splitting of atoms. Large, massive stars eventually explode into supernova (an extremely luminous stellar explosion) due to overheating, whereas smaller stars, like our Sun, eventually run out of energy and become cold, dark, "dwarfs". Astrophysicists can determine when a star was born, how old it is and when it will "die" if they know the distance, mass, magnitude and chemical composition of a star.

Our solar system's Sun formed approximately 5 billion years ago and is expected to have a lifespan of about 10 billion years. Considering that Earth is about 4.6 billion years in age, we have very recently occupied a "middle age" planet, but have very rapidly become the dominant species of this planet. Over the past 4.6 billion years Earth has gone through great unrest. It has been impacted numerous times by extraterrestrial objects, it has experienced plate tectonic collisions to form mountain belts, continents

127

have drifted together and apart, global climate has been cooler than today as well as warmer than today, and Earth has experienced at least 5 major mass extinctions of life.

Our planet is a very dynamic one – one that must be studied from an Earth System Science perspective. Think about the earthquakes and volcanoes generated by plate tectonic movement, the past ice ages, or the demise of the dinosaurs. But, if you are not yet convinced in Gaian philosophy that Earth has the ability to self-regulate its global environment and its inhabitants, just think of the fact that somehow Earth has been a hospitable host for life, of one form or another, for at least the past 3.8 billion years. Self-regulation has allowed Planet Earth to maintain the requirements for life including water, food, and an appropriate climate. If the Earth System gets out of equilibrium it has numerous negative feedbacks to restore equilibrium. By the same token, if the Earth System is perturbed in such a way that there are positive feedbacks it can take it far from equilibrium.

There is little doubt that Planet Earth will exist for another ~ 5 billion years. Because of self-regulation, there is also no question that life of some sort or another will be sustained. Some have, in jest, suggested that if Armageddon does occur that only "cockroaches and seagulls" will survive. The key question, however, is will we (*Homo sapiens*) survive over the long-term?

The only glimmer of hope that I see is to attack the problem at its root source – overpopulation! Many respected scientists and engineers have argued for space colonization or large-scale geo-engineering of Earth as potential solutions to human overpopulation. I reject both of these concepts as untenable. We have already discussed the fundamental limitations of space colonization, and large-scale geo-engineering of our planet to control global temperature is basically "science fiction". Some have argued that we place large reflective surfaces in the atmosphere or in the oceans to keep Earth cool, others have suggested large volumes of aerosols be put in to the atmosphere to increase planetary albedo (reflection of sunlight), while others have recommended "fertilizing" the oceans with iron to stimulate photosynthesis by phytoplankton (small marine algae), as well as carbon sequestration and burial. Can you imagine the cost and unexpected side effects of such large-scale geo-engineering? Even Lovelock himself has proposed Gaian "geophysiological" methods of global geo-engineering as a "stop gap" measure.

I suggest that our global society become one. Considering all the different political and religious ideologies of our world it may well be unrealistic, but in my opinion it is our *only* reasonable option. If our global society becomes one, it would have to consciously begin to reduce human population immediately. Education and contraception would be key components. I would call on the U.N. to assemble the world's best population scientists and sociologist to find a humane way of reducing human population – sort of an analog to the IPCC and climate change, but without the over reliance on subjective computer models and political consensus. *Homo sapiens'* numbers must be reduced to a level that we will be able to adapt to the large environmental changes that are inevitable in our not too distant future – whether it is global warming or the next ice age. If we are small enough in numbers to be adaptable and self-sustainable, yet large enough to perpetuate, we will have a chance at long-term survival as a species. If we are neither - who will be the last two humans on Earth?

VIII) SOURCES OF INFORMATION

Books

Kump, L.R., 2004, *The Earth System (2nd edition)*: New Jersey, Pearson Prentice Hall, 420 pp.

Lovelock, J., 1988, *The Ages of Gaia: A biography of our living Earth*: New York, Norton & Company, 255 pp.

Lovelock, J, 2000, *Homage to Gaia: The life of an independent scientist*: Oxford, UK, Oxford University Press, 428 pp.

Lovelock, J., 2006, *The Revenge of Gaia: Earth's Climate Crisis & The Fate Of Humanity*: UK, Penguin Basic Books, 173 pp.

Lovelock, J., 2009, *The Vanishing Face of Gaia – The Final Warning*: New York, Basic Books, 278 pp.

Mann, M.E., and Kump, L.R., 2009, *Dire Predictions - Understanding Global Warming: The Illustrated Guide to the Findings of the Intergovernmental Panel on Climate Change*: New York, Pearson Education, 208 pp.

Ruddiman, W.F., 2007, *Plows, plagues and petroleum: How humans took control of climate*: New Jersey, Princeton University Press, 202 pp.

Ruddiman, W.F., 2008, *Earth's Climate: Past and Future (2nd edition)*: New York, W.H. Freeman Company, 388 pp.

WEBSITES (In Order Of Use)

1) http://en.wikipedia.org/wiki/The_Population_Bom

b   http://esa.un.org

2) http://www.census.gov

3) http://globalchange.umich.edu

4) http://en.wikipedia.org

5) http://www.soph.uab.edu

6) http://www.cia.gov/library/publications/the-world-factbook/rankoida/2054rank.html

7) http://www.census.gov/population/www/

8) http://www.un.org/esa/population/publications/six billion/sixbill ion.htm

9) http://en.wikipedia.org/wiki/Jane_Goodall

11) http://people.eku.edu/ritchisong/RITCHISO//ENVS CINDTES

12) http://www.globalchange.umich.edu/globalchange2/current/lectures/human_population/humanpop.htl

13) http://www.earth.columbia.edu/events/2005/documents/PDFCohen.pdf

14) http://en.wikipedia.org/wiki/History_of_Technology

15) http://www.britanica.com/EBchecked/topic/1350805/history-of-technology

16) http://en.wikipedia.org/wiki/Anthropocene

17) http://www.geotimes.org/feb08/article.html?id=WebExtra020708html

18) http://www.ipcc.ch

19) http://www.esrl.noaa.gov/gmd/ccgg/trends/

20) http://www.meriamwebster.com/dictionary/Armag eddon

21) http://www.warscholar.com/Timeline.html

22) http://www.hawaii.edu/powerkills/WPP.CHAP16. HTM

23) http://en.wikipedia.org//wiki/war

24) http://en.wikipedia.org/wiki/Atomic_bombings_of_ Hiroshima_nd_Nagasaki

25) http://en.wikipedia.org/wiki/Nuclear_weapon

26) http://www.infoplease.com/ce6/history/A0824719.h tml

27) http://www.infoplease.com/ipa/A0762462.html

28) http://en.wikipedia.org/wiki/List_of_states_with_n uclear_Weapons

29) http://en.wikipedia.org/wiki/Cuban_Missile_Crisis

30) http://www.answers.com/topic/nuclear-winter

31) http://en.wikipedia.org/wiki/Biological_warfare

32) http://www.emedicinehealth.com/biological_warfar e/article_em.htm

33) http://en.wikipedia.org/wiki/Bioterrorism

34) http://www.bt.cdc.gov/bioterrorism/

35) http://www.who.int/topics/bioterrorism/en/

36) http://www.reachingcriticalwill.org/legal/cw/cwind ex.html

37) http://www.emedicinehealth.com/chemical_warfar e/article_en.Htm

38) http://en.wikipedia.org/wiki/Chemical_weapon

39) http://en.wikipedia.org/wiki/Blood_agent

40) http://en.wikipedia.org/wiki/Pulminary_agent

41) http://en.wikipedia.org/wiki/Tear_gas

42) http://en.wikipedia.org/wiki/Blister_agent

43) http://en.wikipedia.org/wiki/Nerve_agent

44) http://en.wikipedia.org/wiki/Incapacitating_agent

45) http://en.wikipedia.org/wiki/Pandemic

46) http://en.wikipedia.org/wiki/AIDS

47)http://www.pandemicflu.gov/general/historicalover view.html

48) http://en.wikipedia.org/wiki/Great_Irish_Famine

49)http://www.doacs.state.fl.us/pi/enpp/ento/dcitri.ht m

50)http://www.freshplaza.com/news_details.asp?id=36 640

51) http://en.wikipedia.org/wiki/Food_crisis

52) http://en.wikipedia.org/wiki/Amartya_Sen

53) http://www.dieoff.org/page40.htm

54) http://en.wikipedia.org/wiki/Famine

55) www.ipcc.ch

56) http://en.wikipedia.org/wiki/Keeling_Curve

57)http://environment.nationalgeographic/globalwar ming/deforestation

58)http://encarta.msn.com/encyclopedia_761586407_5 /fossil_fuels.html

59) http://en.wikipedia.org/wiki/Methane_clathrate

60) http://www.quadrillion.com/numbersystem

61) http://en.wikipedia.org/wiki/Fossil_fuel

62)http://en.wikipedia.org/wiki/Permian_Triassic_extinction_event

63) http://www.scienceagogo.com/news/2004

64)http://www.scienceray.com/Earth-Sciences/Paleontology/The-Great-Mass-Extinction

65)http://www.emc.maricopa.edu/faculty/farabee/BIOK/BioBookpopecol.html

66)http://users.rcn.com/jkimball.ma.ultrviolet/BiologyPages/P/Populations2.html

67)http://www.mansfield.ohiostate.edu/~sabedon/canpl52.htm

68) http://en.wikipedia.org/wiki/Space_colonization

69)http://www.sfgate.com/getoutside/1996/jul/crash.html

70) http://www.tapsns.com

71)http://www.pbs.org/wgbh/evolution/extinction/massert/discuss_08.html

72)http://www.sciencedaily.com/releases/2006/10/061025085208.htm

73) http://en.wikipedia.org/wiki/Extinction

74) http://en.wikipedia.org/wiki/Star

75) http://en.wikipedia.org/wiki/Supernova

76)http://physics.suite101.com/article.cfm/starlifespan

77) http://en.wikipedia.org/wiki/One-child_policy

78)http://geography.about.com/od/populationgeograp
hy/a/onechild.htm

79) http://en.wikipedia.org/wiki/Maya_civilization

80) http://en.wikipedia.org/wiki/David_A_Hodell

81) http://en.wikipedia.org/wiki/Roman_Empire

82)http://en.wikipedia.org/wiki/Decline_of_the_Roma
n_Empire

83) http://en.wikipedia.org/wiki/British_Empire

84)http://en.wikipedia.org/wiki/Evolution_of_the_Brit
ish_Empire

85) http://sfr-21.org/collapse.html

86) http://en.wikipedia.org/wiki/Collapse_(book)

87) http://infoplease.com/ce6/history/A0824719.html

88)http://www.state.gov/www/global/arms/treaties/salt
1.html

NEWSPAPER ARTICLES

Associated Press, 2009, *Report: Some climate change here to stay*: January 27.

Associated Press, 2009, *Airborne emissions up sharply*: February   15.

Boyd, R.S., 2008, *Killer germs resisting world's antibiotics*: McCarthy Newspapers, May 11.

Boyle, R., 2008, *Ark of the Arctic Opens Tuesday*: Fort CollinsColoradoan, February 25.

Eilperin, J., 2009, *Climate Change: We Can Stop the Clock, ButCan't Turn It Back*: Washington Post, February 2.

Lacey, M., 2008, *The Global Food Crisis*: New York Times Service, April 20.

Lee, J.H., 2009, *North Korea warns of war*: Associated Press, March 10.

News Service Reports, 2009, *China to aid farmers hit by worst drought in 50 years*: February 8.

Perlman, D., 2009, *Experts: Fading forests linked to warming*: San Francisco Chronicle, January 24.

Schmid, R.E., 2008, *We're Wiping Out Our Closest Relatives*: Associated Press, October 13.

Schmid, R.E., 2009, *How Serotonin Turns Solitary Locusts Into A Swarm*: Associated Press, February 9.

Stringer, D., 2008, *'Tsunami' of Hunger*: Associated Press, April   23.

Szabo, L., 2009, *Cleaner air may be adding three years to your life*: USA TODAY, January 22.

Will, G., 2009, *Global cooling was hot topic, too*: Syracuse Post-Standard, February 15.

Zhuang, Q., et al., 2009. *Global methane emissions from wetlands, rice paddies, and lakes*: EOS, American   Geophysical Union, February 3.

PHOTO CREDITS

1) J - Curve: www.chrismartenson.com

2) A- Bomb: U.S. Government Image